# Adam Smith and His Legacy for Modern Capitalism

# Adam Smith and His Legacy for Modern Capitalism

PATRICIA H. WERHANE

New York Oxford
OXFORD UNIVERSITY PRESS
1991

Oxford University Press

Oxford New York Toronto
Delhi Bombay Calcutta Madras Karachi
Petaling Jaya Singapore Hong Kong Tokyo
Nairobi Dar es Salaam Cape Town
Melbourne Auckland

and associated companies in
Berlin Ibadan



Published by Oxford University Press, Inc.
200 Madison Avenue, New York, NY 10016



Library of Congress Cataloging-in-Publication Data
Werhane, Patricia Hogue.
Adam Smith and his legacy for modern capitalism / Patricia Werhane.
p. cm. Includes bibliographical references and index.
ISBN 0-19-506828-9
1. Smith, Adam, 1723-1790. 2. Capitalism.
3. Economics—Great Britain—History—18th century. I. Title.
HB103.S6W38 1991 330.15′3—dc20
90-20791

1 3 5 7 9 8 6 4 2

Printed in the United States of America
on acid-free paper

*For Chuck*

# PREFACE

Adam Smith is one of the mostly widely read eighteenth-century thinkers, enjoying a scholarly reputation among economists, social scientists, political theorists, as well as philosophers. Yet Smith completed only two books, *The Theory of Moral Sentiments* and the *Wealth of Nations*, and he is best known for three or four quotations from the latter text. Taken out of context, these quotations appear to argue that human beings are motivated primarily by self-interest. In economic affairs such motivation is justifiable because in a competitive environment, self-interests usually counteract one another, to the economic dissatisfaction of no one. Indeed, under optimal competitive conditions, the market, what Smith calls the "invisible hand," actually increases the public good by improving the economic well-being even of those not directly competing in the marketplace, despite the primarily self-interested intentions of those participating in economic affairs. Although this simple and fairly straightforward argument has been attacked by a number of commentators on Smith, the perception that this represents the heart of Smith's basic theses in the *Wealth of Nations* persists in many quarters, particularly in the more popular readings of Smith's works. Moreover, versions of this interpretation have been unduly influential in analyses of the foundations and justifications of economic capitalism.

My purpose in writing this book is to dispel the conclusion that Adam Smith was in any sense an egoist, that self-interest is or should be the motivating norm of a political economy, or that the invisible hand is the intentional perpetrator of economic well-being. Smith's sensitivity to the diversity of human motivation in the *Theory of Moral Sentiments*; his focus on rights evidenced in student notes taken from his lectures, now reproduced in the *Lec-*

*tures in Jurisprudence*; his critique of business people, business, and corporations in the *Wealth of Nations*; and the theme of justice that runs through all his works belie a simple interpretation of his views. It is true that Smith argues that pursuing one's private interests need not conflict with, and indeed may contribute to, the public good, but only under specific conditions in which economic liberty operates in the context of prudence, cooperation, a level playing field of competition, and within a well-defined framework of justice.

I shall reexamine Smith's major works in light of new commentaries on these texts, concentrating on some of the crucial interpretative confusions that repeatedly bother readers of Smith's works. Chapters 1 through 3 will focus on a close reading of the *Theory of Moral Sentiments*, the *Lectures on Jurisprudence*, and the *Wealth of Nations*. In brief, I shall argue that the social as well as the selfish passions are equally motivating even in economic affairs, that justice and not benevolence is the basic virtue in all three texts and plays a central role in all human affairs, and that the "invisible hand" about which so much has been written is not the force that drives Smith's ideal political economy. Chapter 4 concerns itself with the question of Smith's alleged individualism and discusses the role of societal institutions in Smith's thought. Chapter 5 analyzes Smith's labor theory of value, the relation of labor to property, and the question of equality to which Smith sometimes confusingly refers. Chapter 6 lays out Smith's ideal political economy and responds to some of his critics. The success or failure of the book will lie in whether I can convince the reader of Smith's originality as both a moral philosopher and a political economist, of his sensitivity to the frailty of human nature, and of his success in drawing together spheres of morality, justice, and economics in his development of the norm of a political economy.

The first draft of this book was written in 1988–1989 during a Rockefeller Fellowship at the Institute for the Study of Applied and Professional Ethics at Dartmouth College. The generosity of the Rockefeller Foundation and Dartmouth College for this fellowship, the numerous colleagues and friends at Dartmouth who assisted in this project, the staff at Baker Library, and the executive director of the Ethics Institute, Deni Elliott, made this volume possible. The project was also made possible by a sabbatical leave and financial support from Loyola University of Chicago, a National Endowment for the Humanities–Loyola Faculty Development Grant, and sustained assistance from Research Services, the Department of Philosophy, and the administration at Loyola. I want to thank all these institutions for their invaluable support.

Some time ago I was introduced by my former colleague Tom Donaldson to Adam Smith and the problem of mercantilism. Subsequently, Robert Hanneford invited me to speak at Ripon College on the role of Adam Smith's philosophical legacy for contemporary problems in economics and ethics. That symposium promoted my conviction that Adam Smith had not been always correctly interpreted, at least in the more popular literature on his philosophy and economics, and so the project for this book was initiated.

A number of people have commented on parts of the manuscript, including Norman Bowie, Robbin Derry, Bernard Gert, Ronald Green, Charles Griswald, David Kirk Hart, Harriet Hogue, David Levy, Amartya Sen, Robert Solomon, Mark Waymack, Paul Wendt, and some anonymous reviewers of the text. I am most grateful for this assistance, but I am responsible for transposing their comments into the reality of this book. Ann Dolinko read and corrected the quotations and footnotes, a critical and thankless task, Carmela Epright assisted with the bibliography, and Cynthia Rudolph was responsible for other secretarial assistance. A number of family members and friends provided spiritual encouragement while this project was under way, for which I am most grateful.

*Chicago* P. W.
*December 1990*

# ACKNOWLEDGMENTS

A much abbreviated version of the Introduction and Chapters 1 and 3 appeared under the title "The Role of Self-Interest in Adam Smith's *Wealth of Nations*" in the *Journal of Philosophy*, 86, November 1989. These sections are reprinted with permission of the *Journal of Philosophy*, Inc. Chapter 4 originally appeared under the title "Freedom, Commodification, and the Alienation of Labor in Adam Smith's *Wealth of Nations*" in *Philosophical Forum*, 1991 (forthcoming), and is reproduced with the permission of that journal.

# CONTENTS

Adam Smith and His Legacy for Modern Capitalism

# Introduction

Adam Smith has been called a major Western philosopher, the greatest political economist, the father of the Industrial Revolution, and even the founder of social science. Whether or not it is true, as Joseph Schumpeter claims, that "the *Wealth of Nations* does not contain a single analytic idea, principle, or method that was entirely new in 1776,"[1] Smith is, nevertheless, justly famous. In the *Wealth of Nations* he brings together disparate eighteenth-century economic concepts and theories into a unified theory of political economy. In critiquing European economic practices of his day and by organizing and ordering economic theory, he not only creates economics as a distinct intellectual discipline and body of knowledge, but more importantly, he provides a coherent theory, now called *classical economics*, that has served for over two centuries as an essential component in the development of economic thought.

Yet almost since his death there has developed a number of caricatures of Smith's well-known book the *Wealth of Nations* (*WN*), a series of interpretations that do not always accurately represent the content and spirit of the text. Although these interpretations are not accepted among all serious scholars of Smith and although no one thinker holds all of these views, as a collection of commentaries they have influenced contemporary sociological, economic, and ethical analyses of the *WN*. The most prevalent element of this series of interpretations is summarized in Albert Hirschman's comment that "the main impact of *The Wealth of*

*Nations* was to establish a powerful economic justification for the untrammeled pursuit of individual self-interest."[2]

In the seventeenth century Thomas Hobbes argued that deprived of a societal framework that included law and order, each of us in a "state of nature" would be motivated to seek our own self-preservation in disregard of others. Because of this argument Hobbes is usually depicted, probably erroneously, as a psychological egoist alleged to hold the view that we are always motivated by our selfish passions and self-interests. This Hobbesian picture of human motivation has played an influential and, in some schools, even a dominating role in ethics and economics. The question then becomes, how do we restrain, control, and direct our egoistic passions to avoid harm, protect long-term self-interests, and promote the public good?

Adam Smith is sometimes read as having promulgated this egoistic picture of human motivation and as having solved the problem of the dichotomy between natural selfish passions and public interests. Smith is said to have adopted a thesis attributed to a contemporary, Bernard Mandeville, if not to Hobbes, the notion that human beings are solely motivated by selfish passions. Even what appear to be benevolent acts are derived from selfish motives, according to Mandeville, and we help others only when there is some gain for ourselves. Fortunately for humankind, our desire to be admired and praised, coupled with restraints created by competing self-interests, turn out to be in the public benefit, at least as that concerns economic welfare. Accepting the view current at the time, that by and large, selfishness was evil and benevolence was virtuous, Mandeville concluded that the private vices of selfishness and greed could lead to public benefits.[3]

Smith's "Mandevillean" picture of human motivation is usually traced to his statement in the *WN*:

> It is not from the benevolence of the butcher, the brewer, or the baker, that we expect our dinner, but from their regard to their own interest. We address ourselves, not to their humanity but to their self-love, and never talk to them of our own necessities but of their advantages.[4]

When one reads this passage out of context with the rest of the text and in isolation from Smith's other writings, it appears that at least in economic affairs, one usually acts in one's unrestrained self-interest. Linking that passage with Smith's statement that an economic actor who "intends only his own gain . . . is . . . led by an invisible hand to promote an end which was no part of his intention,"[5] some commentators attribute to Smith the conclusion that acting selfishly, at least in the marketplace, is all right in an unregulated economy, because competitive forces of other equally self-interested actors will balance, and indeed control, these acquisitive activities. In fact, according to this reading, the balance of self-interests in a competitive efficient market usually improves the economic well-being even for those not directly participating in market transactions.[6] In Smith's own words, when "all systems either of preference or of restraint . . . being thus completely taken away, the obvious and simple system of natural liberty establishes itself of its own accord."[7] Thus, as some students of the *WN* conclude, Smith means to imply that the harmony of these individual pursuits will, autonomously and unintended by the actors, often produce economic good.

The Mandevillean thesis links all motivation with the selfish passions, but more careful readers question this as Smith's view. Rather, they argue, "self-interest and not selfishness is the very foundation of his [Smith's] edifice of thought; it is the important driving force in real life, ethically positive and of social benefit under definite conditions."[8] That is, the butcher and baker are not necessarily selfish but simply are not concerned with the interests of others, at least in economic affairs, except when those interests affect their business concerns. Thus it is self-interest, not selfishness, that is the dominant motive, at least in economic matters, and under the proper conditions this self-interest can produce social benefits.

The further thesis that an unregulated market is impartial and self-correcting comes from these interpretations of Smith. An unregulated market is impartial and self-correcting just because the collection of self-interested economic actors competing with one another to satisfy consumer demand creates a self-constraining system of which no one actor is allowed to take advantage for very long. "The pursuit of rational self-interest is an unmitigated social good,"[9] according to this view. A self-correcting, impartial, consumer-driven market needs no further interference because it protects the self-interests of individual economic actors, responds to consumer demands, and promotes well-being and economic prosperity impersonally without bias toward any one group of individuals.

These interpretations of Adam Smith, made in the nineteenth century, have been challenged by a number of contemporary scholars.[10] Nevertheless, such views persist and are still called by some the "prevailing view."[11] They are seen as a collection of views and have influenced sociological and economic analyses of *WN* particularly because they discuss topics as diverse as human motivation, socialization and social organizations, rational choice theories, utility and utility maximization, foundations of free enterprise, and the justifications of an unregulated market system. For example, the contemporary social scientist Robert Frank declares that "the modern behavioral scientist's focus on self-interest traces directly back to Adam Smith."[12] And in his new book on socioeconomics Amitai Etzioni declares: "At the foundation of the neoclassical paradigm [the thesis that "individuals . . . maximize their utility, rationally choosing the best means to serve their goals"[13]] is Adam Smith's *Wealth of Nations* view: that self-serving people in a competitive market provide for the most efficient mode of social organization, especially of economic activities."[14]

Such interpretations of the *WN* also form the philosophical basis of some contemporary economic and political theories, in particular the Chicago school represented by George Stigler and Milton Friedman, and rational choice theory represented in views such as those of Anthony Downs. Sometimes, too, such theses are attributed to neoclassical economic theory, although there are many in this school who make counterclaims to such readings of Smith. Although it is true that no one economist holds all the theses attributed to this caricature of Smith, various thinkers hold variations of these positions. These theses include the following: First, "man is eternally a utility maximizer";[15] that is, at least in making rational choices, we seek to maximize utility. Second, some (but not all) theorists contend that we are ordinarily narrowly self-interested. Others assert that we are merely "non-tuists," unconcerned with the interests of others or unconcerned with the interests of others at least in regard to economic exchanges.[16] Third, human beings are "social atoms";[17] that is, either we are radically individualistic, or at least, individuals are the basic units of motivation, preference, decision making, and organization.[18] Fourth, it is sometimes claimed that under Smith's invisible hand, a perfectly competitive market is "a morally free zone";[19] that is, "the market satisfies the ideal of moral anarchy."[20] Others see the perfectly competitive market as the "exemplar of rational morality."[21] Smith is often considered the father of these schools of thought by both their proponents and their critics.

In this book I shall conduct a sustained argument opposing these sets of interpretations of the *WN*. Although many of these theses may be valid in their own right—a validity that I shall not question in this book—they are erroneously attributed to Smith and thus must be justified on other grounds. For the sake of simplicity, I shall refer to these series of commentaries as the *self-interest views*; because even though various individual sources pick up only certain aspects of this series of interpretations, almost every source I shall criticize views self-interest as the dominating motive in the *WN*. I shall challenge that thesis. Moreover, I shall argue, Smith indeed develops a model political economy, but it is not the economic model I just outlined. Indeed, his analysis questions many of the theses that are the core of these ideas, in particular, that human beings are primarily self-interested rational utility maximizers, that we are social atoms, and that a perfectly competitive market is either "free from morality" or an "exemplar of rational morality." While being an early advocate of what eighteenth-century French economists called *laissez-faire*, the pivotal roles of self-restraint, cooperation—what Smith calls "equality of treatment"—and justice form a model framework for economic activity. Such a model, driven by justice as well as utility, questions all but the most trivial forms of egoism and market anarchy as ideals in a commercial political economy. How did these readings of the *WN* develop, and what are their difficulties?

## The Adam Smith Problem

Adam Smith's first book, the *Theory of Moral Sentiments* (*TMS*), is a treatise on moral psychology, and his second and best-known work is the *Wealth of Nations* (*WN*). In addition, there are in existence two collaborative sets of student notes from Smith's lectures on justice gathered together in what is called the *Lectures of Jurisprudence* (*LJ*). Although the *TMS* was immensely popular during Smith's lifetime, this text has not enjoyed the widespread continued reading afforded to the *WN*. As a book on moral psychology, it is overshadowed by the works of other great seventeenth- and eighteenth-century moral philosophers, for example, Hobbes, Shaftesbury, Hutcheson, and Hume, each of whom places his moral philosophy in a larger epistemological framework, a framework scarcely developed by Smith in any of his writings.

In the *TMS*, Smith argues that human beings are motivated by selfish, social, and unsocial passions; that our interests are not merely self-directed; that justice as well as self-command, benevolence, and prudence are moral virtues; and that the phenomenon of sympathy plays a central role in approbation and moral judgments. As a political and economic treatise criticizing the status quo and defending an unregulated industrial political economy as a means for economic growth, the *WN* was a revolutionary work in the eighteenth century, a work whose influence was and still is enormous. It is no surprise, then, that the *TMS* receives less attention. It is clear that the *TMS* and the *WN* are obviously quite different works, and on first reading it appears that there are serious philosophical differences between the two texts. For example, in the *WN*, if self-interest, however defined, is the driving force in economic relationships, a force that nevertheless produces welcome consequences, this conclusion would appear to contradict the arguments in the *TMS*. One interpreter suggested that "in the *Moral Sentiments*, he [Adam Smith] ascribes our actions to sympathy; in his *Wealth of Nations*, he ascribes them to selfishness. A short view of these two works will prove the existence of this fundamental difference."[22] Therefore, a number of nineteenth-century commentators declared that there was what came to be known as the "Adam Smith Problem." They contended that the *WN* and the *TMS* were not merely inconsistent, but worse, they even contradicted each other in regard to the sources of human motivation, human conduct, and virtue. Some commentators identified self-interest with selfishness as the basic principle of motivation and norm of human conduct in the *WN* and concluded that this was contradictory to the notions of sympathy and benevolence expounded in the *TMS*.[23] Even if individual self-interest rather than selfishness were the prevailing principle of motivation in the *WN* and underlay the concept of the invisible hand, this emphasis on individualism, it was argued, would be inconsistent with Smith's depiction of the individual as a social being when sympathy is part of the social order, a theme that dominates the *TMS*.

Careful scholarship beginning in the late nineteenth century has somewhat cleared up the alleged contradictions between the *TMS* and the *WN*. Merely from a historical perspective, it is difficult to imagine that Adam Smith did not have the *TMS* in mind when he wrote the *WN*. The *TMS* precedes the *WN* chronologically, and he later revised the *TMS* a number of times after he finished the *WN*. In the last revision of the *TMS* he added the virtue of self-command, a notion that some commentators regard as implicit in the *WN*.[24] Therefore, at least from a historical point of view, although the subject matter of the *TMS* and the *WN* is different, it is unlikely that Smith meant to develop radically inconsistent ideas in these two works. Rather, in the *TMS* Smith focuses on morality, motivation for moral behavior, and moral judgments, whereas in the *WN* he concentrates on economic activities and economic behavior. The difference in subject matter, then, should account in part for the alleged inconsistencies.[25]

Despite the historical resolution of the Adam Smith Problem, even today some readers, primarily philosophers, concentrate on the *TMS*, assuming that the *WN* has little to do with moral philosophy. Others, many of whom are social scientists, including political theorists and economists, focus almost exclusively on Smith's second work, the *Wealth of Nations* (*WN*). Fewer refer to his earlier treatise on moral psychology, despite historical evidence to suggest that he meant the *TMS* to form the moral foundation for his other writings. When applying Smith's views to modern economic theory, commentators do not always take into account the moral psychology of the *TMS* from which the notions of self-interest, natural liberty, and the invisible hand are derived.

Moreover, the well-known Smith scholar, Jacob Viner, who is intimately familiar with both works, argues that some differences between the *TMS* and the *WN* cannot be resolved and suggests that attempts to complete reconciliation distort one or another of the texts. According to Viner, by breaking with some of the ideas of the *TMS*, the *WN* is a more mature and, indeed, better text.[26] In particular, Viner finds that the optimistic presentation of the harmonious natural order emulated in the moral sentiments of human beings, as Smith depicts it in the *TMS*, is less evident in the

*WN*. There Smith is more of a realist. Although he appears to develop the ideal of an economic order that is in harmony with the natural order, "Smith acknowledged exceptions to the doctrine of a natural harmony in the economic order even when left to take its natural course."[27] Second, the notions of sympathy and benevolence are virtually absent in Smith's description of economic affairs in which the passion of self-interest and the anonymity or distance that accompanies economic relationships dominates its depiction. Third, Viner notices that in the *TMS* Smith's analysis of what he takes to be the natural order underlying moral sentiments of diverse peoples argues for a theory of moral psychology that is both universally descriptive of human nature and static. Thus basic human nature does not change or evolve. This view is not atypical of pre-Darwinian thinking, but it is in contrast with the evolution of history, law, society, and political economy that Smith elaborates in the *LJ* and the *WN*.[28]

The question of how to read the *TMS* and the *WN* as consistent texts remains a serious issue in Smithian scholarship. The most notorious concept for interpretative resolution is that of self-interest, but there are other difficulties of interpretation as well. Apparent discrepancies between a number of seminal notions have misled commentators either to find strong differences between the two texts or nonexistent parallels. Some of these confusions include the following.

The *WN* allegedly contends that rational economic actors act in their own self-interests to maximize their own well-being. This interpretation of the *WN* results in the conclusion that the notion of self-interest in the *WN* is different from that in the *TMS*. In the chapters to follow I shall question this narrow interpretation of the *WN*. But given that reading of the *WN*, students of Smith often differently define self-interest in that text. Many feel that in the *TMS* self-interest should be read as enlightened self-love linked to benevolence and that in the *WN* self-interest should be identified with selfishness or even greed.[29] Others read Smith as a "non-tuist" in his definition of self-interest; that is, one does not take an interest in the interests of others in economic affairs. In an attempt to resolve apparent differences between the *TMS* and the *WN*, at least one commentator, Joseph Cropsey, contends that the notion of self-interest in both works is Hobbesian. Despite their varying readings of the texts, commentators almost always focus on self-interest and ignore the role of the social passions. In the *TMS* the social passions are motivating forces equal to the selfish passions, whereas in the *WN* the social passions are the sources of cooperation and coordination without which no economy could operate. Therefore, in both works—as I shall argue at some length—the social passions and interests are as important as is self-interest. It is an error to label Smith an egoist, and it is not the case that self-interest is the "granite of the *Wealth of Nations*,"[30] as the economist George Stigler named it. Moreover, although self-interest may appear to be non-tuism, non-tuism does not account for the totality of human motivation even in economic exchanges.

A second kind of misreading has to do with Smith's treatment of passions and interests. According to A. O. Hirschman, in both texts Smith conflates passions with interest. Furthermore, the "noneconomic passions," those having little to do with self-interest, become parasites of the economic desires, so that there is no contrapositioning balance among different conflicting passions.[31] A careful analysis of the *TMS* and the *WN*, however, does not support Hirschman's conclusion, as in both texts Smith makes a number of distinctions between the passions and various kinds of interests. In particular, Smith distinguishes between the selfish passions that are directed toward the self and self-interest that can be either virtuous or evil.

Accompanying these readings of Smith are other questionable analyses of important concepts. Glenn Morrow suggests that "the virtue of benevolence" dominates the *TMS*, whereas "the virtue of self-interest regulated by justice" governs economic relationships in the *WN*.[32] However, justice, not benevolence, is the basic virtue in both works. Smith has been misidentified as a utilitarian; yet he repeatedly attacks utilitarianism in the *TMS*. In the *WN* he assumes that a viable political economy will provide well-being for its citizens, but this is not its only role, because equally as importantly, an ideal political economy provides the conditions for economic liberty, the protection of perfect rights, and the framework of justice.

Because of the role of natural liberty in Smith's ideal political economy, historically there is often insufficient attention to Smith's (albeit obscure), theory of justice, a central principle in the *TMS*, the *LJ*, and the *WN*. This oversight has been remedied in recent works by Knud Haakonssen, Istvan Hont, and Richard Teichgraeber. In emphasizing the importance of economic liberty in an unregulated market, Smith's attention to the importance of individual pursuit of economic interests is still often read as championing an almost anarchical state of economic exchanges in a vacuum of regulatory constraints. Such a reading ignores the fact that Smith was an eighteenth-century thinker who presupposed that a society can function only within an institutional framework of constitutional law, order, and justice that has a strong societal moral and religious underpinning.

According to Smith, justice is both a negative principle that protects persons and their rights and property from harm and a positive principle of fair play. In the *TMS* justice is a virtue, an internalized ideal of fair play, and "the main pillar that upholds the whole edifice [of human society]."[33] Linking justice to perfect rights in the *LJ*, Smith argues that natural jurisprudence is properly restricted to commutative justice embodying duties to protect perfect rights. This connection allows Smith to argue that justice, but not benevolence, is a virtue of economic affairs. In the *WN* justice takes on the role of those laws necessary to protect persons and their rights and properties from harm and to enforce fair contracts, debts, and payments, laws that reflect, at least imperfectly, natural jurisprudence.[34] Still, at least a few commentators simply ignore the role of justice in the *WN*.[35]

Finally, there is often an overemphasis on the notion of the invisible hand, a term mentioned exactly once in the *TMS* and once in the *WN*. Sometimes the alleged role of the impartial market is depicted as, for example, "the ultimate governor which controls the self-love of individuals . . . the ultimate natural harmony of individual interests."[36] This exaggerated picture of the invisible hand unduly personifies it. For Smith the invisible hand is a result of economic interchanges, not the engine that drives these exchanges. Thus how the invisible hand "regulates" the economy depends very much on the societal framework in which market activities are conducted and the kinds of exchanges that contribute to it.

In summary, apparent contrasts between the *TMS* and the *WN*, coupled with what I have called the self-interest interpretations of the *WN*, in their various formulations, raise a number of difficult issues. The apparent focus on egoism or self-interest as the driving force in economic activities in the *WN* appears to contrast sharply with Smith's development of sympathy or fellow feeling as central to morality in the *TMS*. The theme of individualism, supposedly pervasive in the *WN*, is said to differ from the focus on human beings as social beings in the *TMS*. The idea of natural liberty developed in the *WN* seems to be in direct contrast with self-command and justice in that work. Did Smith mean to write two separate treatises? Is he implying that morality and a political economy are unrelated or that the principles governing economic relationships are different from those governing morality? Is the *WN* an advance of Smith's thinking, or is the apparent egoistic individualism in the *WN* a misreading of that text?[37]

In light of these difficulties of interpretation, I propose to reexamine the *WN* in light of Smith's moral psychology developed in the *TMS* and his treatment of rights, property, and jurisprudence in the *LJ*. I shall analyze the notions of self-interest, the social passions, sympathy, justice, the competitive market, and the invisible hand in these works. Without demanding a radical revision of one's thinking about Smith's philosophy of economics, I shall argue that Adam Smith was a consistent as well as a brilliant thinker whose views in the *TMS* were meant to set the stage for the analysis in the *WN*. Smith's distinctive contribution to moral psychology is his unique challenge to egoism and altruism. Smith does not set up a dichotomy between these two views. Rather, in the *TMS* he criticizes any moral theory that derives its basis for moral judgments merely from self-interest and, equally, questions any moral theory that derives these judgments solely from benevolence. Distinguishing passions from interests, Smith argues that human beings are not motivated merely by selfish passions, that both prudence and benevolence are virtues of the self-directed and social interests, and that the basic virtue is justice. In the *TMS* Smith lays the groundwork for his development of a political economy, an economy not driven merely by self-interest in which cooperation within the institutional framework of justice provides other conditions for a viable economy.

If this reading is accurate, it will require at least a rethinking of the notion of self-interest in the *WN* in light of the equal importance of the social passions, a judgment about the extent to which utility is a good, a taking into account of Smith's negative theory of justice in economic activities, and a reexamination of the scope and the role of an impartial market in economic life. Smith's treatment of the passions and interests in the *TMS* suggests a reading of his notion of self-interest in the *WN* as restrained "self-command" working in conjunction with, but not dominating, the social passions. Smith's notion of justice, though surely not adequately developed, is central to the *WN* and must be accounted for in the context in which natural liberty and the invisible hand operate.

I shall argue that the so-called self-interest views are misreadings of what Smith meant in the *WN*. A neo-Hobbesian reading of Smith's notion of self-interest is wrongly focused, and the emphasis on the primacy of individual self-interest does not adequately account for the equally important roles of the social interests and justice in the *WN*. Smith did indeed present a solution to the problem of how one controls private passions in the public interest, but his solution is not the one ordinarily attributed to him. We shall see how a proper reading of self-interest in the context of Smith's claim that human beings are motivated by both selfish and social passions responds to Hobbes and affects Smith's seminal notion of the invisible hand and thus his theory of a free market.

Smith's analysis favors a commercial economy with a paucity of economic regulations as the best means to achieve efficiency and economic growth. But although it appears that his notion of natural liberty is a "license" to do as one pleases restrained only by the competitive actions of others, this interpretation ignores his emphasis on the importance of jurisprudence in economic affairs. Free-market exchanges can unintentionally produce economic well-being, but only under certain specifiable conditions. I shall argue that the character of a free market is only as impartial as are those economic persons acting on its behalf. The invisible hand is merely a description of the working of a free market as a result of economic interchanges, not something more—the "ultimate governor" or impartial spectator. Smith's thesis is that the invisible hand works because, and only when, people operate with restrained self-interest in cooperation with others under the precepts of justice. To operate otherwise when that action causes harm to others would be a "violation of fair play."[38] Smith's negative theory of justice, his claim that "there can be no proper motive for hurting our neighbour,"[39] plays a more important role in the *WN* than is sometimes thought.

## New Adam Smith Problems

In addition to the original Adam Smith Problem, in this century there have developed what I would call *New Adam Smith Problems*. First, a number of commentators have suggested that the method and the philosophy of the *WN* and the *TMS* are identical, thereby translating central terms of the *TMS*—in particular, the notions of sympathy and the impartial spectator—into economic roles in the *WN*. Second, some readers apply contemporary sociological concepts to Smith's writings. Thus he is often wrongly identified as either a methodological individualist or a methodological collectivist. Third, scholars such as Viner, by finding discrepancies between the two works, sometimes overemphasize their incongruities.

The notion of sympathy is a central concept in the *TMS*, but this term does not appear in the *WN*. Nevertheless, both P. L. Danner and Robert Boynton Lamb claim that the notion of sympathy, as well as our desire for approval by others, governs property and exchange relationships, particularly as they are treated in the *WN*. According to Danner, sympathy in the *TMS* translates in economic terms into "exchangeable value" in the *WN*,[40] and Lamb feels that "'sympathy' rather than self-interest is the basis of property in Smith's system."[41] But Danner and Lamb read too much into the *WN*. It is true that Smith does not think that self-interests are the only bases for exchangeable value or property arrangements and that both property arrangements and exchangeable value depend on social as well as selfish interests. However, it is unlikely that Smith meant sympathy to be identified either with exchangeable value or as a basis for property. The reason is that sympathy is neither a sentiment nor a passion, but rather, in Smith's terminology, sympathy is a means through which we understand, but never fully experience or empathize with, the feelings and passions of others. Thus sympathy is not the source of our desire for approval, and so it does not play the kind of role in the *WN* that Danner and Lamb imagine.

Some commentators reconcile the *WN* and the *TMS* by going so far as to argue that the concept of the impartial spectator that Smith develops in the *TMS* translates into the free market or the invisible hand in the *WN*. In the *TMS* Smith's impartial spectator is the average person disinterestedly disengaging himself or herself from a particular situation in order to examine and approve or disapprove of a motive or action and thus make a moral judgment and set standards for moral rules. Keeping in mind this notion, A. L. Macfie argues, "The impartial spectator in fact makes no appearance in the *Wealth of Nations*. He there becomes the impersonal market."[42] This has been interpreted to mean that "the free market is the universal reign of the impartial spectator, the invisible but rational hand, which rational producers for their self-interest must obey."[43] This reading, however, depicts the invisible hand as an autonomous regulator of human affairs. Because the invisible hand is a dependent rather than an independent variable in economic affairs, Smith does not use the notion of the impartial spectator in this context, and it is unlikely that he would find this interpretation appealing.

A second set of Adam Smith Problems has developed in contemporary analyses of the *TMS* and the *WN*. Sometimes coupled with reading Smith as an egoist and sometimes independently, Smith is regarded either as a radical individualist or, conversely, as a collectivist. For example, while contending that Smith is a collectivist in the *TMS*, Glenn Morrow suggests that in the *Wealth of Nations* "Adam Smith conceives of the economic order as purely a collection of competing individuals."[44] Alternatively, Warren Samuels claims that "Smith . . . blended methodological individualist and methodological collectivist levels of analysis."[45] Smith's focus on institutions and social interactions in the *TMS* and the *WN* is often ignored and must be reckoned with, but one must be careful not to read into it a later concept of collectivism. Smith is clearly an eighteenth-century individualist. But the theme of radical individualism is not in keeping with his analysis of a political economy, either. Thus one must give due consideration to Samuels's statement that according to Smith "institutions . . . govern the answer to the question of whose liberty is to be achieved."[46]

Third, the differences that Viner finds between the *TMS* and the *WN* are not as dramatic as he sometimes indicates. The absence of benevolence and sympathy in economic affairs does not demonstrate a discrepancy between the *TMS* and the *WN*. As we shall see, it is justice, not benevolence, that is the basic virtue in the *TMS*, and it is justice, too, that is a necessary condition for a well-functioning economy. Although it is true that Smith's account of the natural order contributes to a static view of human nature, that same natural order is the basis for his more evolutionary analysis of morality and the historical development of political economies. So it may be the case that Smith is saying that there are general principles or a natural order underlying all human life. More will be said about the new Adam Smith Problems in the following chapters.

## Textual Difficulties in the *Wealth of Nations*

In addition to the problems raised in interpreting Smith's texts, in the *WN* Smith himself creates difficulties for even the most cautious reader. He is sometimes unclear about the definition and development of some crucial notions. This is particularly noteworthy in his treatment of justice, property, labor, the division of labor, and what he calls "equality of treatment." Moreover, Smith sometimes describes actual economic phenomena, but in other places he presents a normative analysis. He seldom distinguishes these two modes of discourse, so that one is never sure whether his analysis is a description, a condemnation, a recommendation, or a prescription for what should be the case.

In analyzing justice Smith sometimes uses the term *natural jurisprudence*. In other passages he adopts a more Humean notion of justice, depicting laws of justice as conventions, and considers their utility in social, political, and economic exchanges. Thus Smith appears to combine aspects of an earlier natural law tradition with Hume's more conventional view of justice.

Smith asserts that in a workable political economy, "the difference of natural talents in different men is, in reality, much less than we are aware of,"[47] that we have obligations to the poor, and that "justice and equality of treatment . . . the sovereign owes to all the different orders of his subjects."[48] At the same time it is patently evident that property and property rights are basic to Smith's idea of a well-functioning economy, although one must turn to the *LJ* to get a proper understanding of his views of these notions. Property is both an adventitious and a perfect right, a questionable conflation of two quite different qualifications. Moreover, in the *LJ* as well as in the *WN* Smith argues that property gives rise to civil government, that property creates inequalities, and that one of the tasks of civil government is protect the rich against the poor. But how can civil society both protect property owners from the propertyless poor and achieve equality of treatment as well? The explanation of his resolution of this dilemma will be one of the tasks of this book.

A similar problem arises in Smith's analysis of the division of labor. I think it is not an exaggeration to say that the division of labor is the most important economic concept in the *WN*. It, along with stock and rent, are the source of new wealth, economic growth, and therefore economic well-being in any economy. About the division of labor Smith writes, "the greatest improvement in the productive powers of labour, and the greater part of the skill, dexterity, and judgement with which it is any where directed, or applied, seem to have been the effects of the division of labour."[49] In Book I of the *WN* he extols the virtues of the division of labor, but in Book V he declares that overspecialization "corrupts the courage of [the labourer's] mind . . . [and] even the activity of his body."[50] Smith is an advocate of high wages, the abolition of apprenticeships, and the free movement of labor. Yet he finds that employers often collude with one another to keep wages low, that laborers are dependent on their employers, and that high wages sometimes have negative effects on profits and thus on the accumulation of capital for reinvestment. What, then, is Smith's position on labor and the rights of workers?

Smith's insights into the possible evils of industrialization just cited and his historical analysis of economic development have led some commentators to see Smith as the intellectual predecessor to Karl Marx and to interpret him as much more sympathetic to a Marxist point of view than is ordinarily thought.[51] However, as I shall argue in detail, Smith's use of history is different from Marx's, because Smith would contend that his own stages of economic development are neither a sequential nor an exhaustive description of economic evolution. More importantly, Smith's labor theory of value, which identifies productivity as a commodity, serves to liberate the laborer from her productivity. This distinction of the laborer from her productivity is not merely abstract freedom, Smith would point out, because such a division allows the laborer to barter her productivity without selling herself. It is the identification of the laborer, laboring, and productivity, not its separation, that Smith finds morally questionable. Thus Smith may be read as presenting an anticipatory response to Marx's theory of alienation. Although Marx was aware of Smith's method and model and found them lacking, it is safe to say that Smith would have thought his own theory adequate to respond to Marx.

Perhaps overoptimistically, Smith finds that the laborers' independence in a commercial context in which productivity is valued as a commodity allows the laborer to compete, barter her productivity, and thus improve her well-being. Under the conditions of perfect competition, competition among property owners restrains their power while creating a climate of full employment and thus competition for labor. Moreover, Smith thought that the ideal of universal public education would free laborers from the corruption of their identification with their work. Thus although economic equality is never achieved, nor is this a goal for Smith, a free competitive market has a "leveling" effect on power and an enhancing effect on both the freedom and economic well-being of labor.

## Smith's Ideal

Smith is allegedly a champion of industrialized laissez-faire economics. In Book IV of the *WN* Smith conducts a sustained critique of the mercantile practices of the British government of his day that regulated domestic and foreign trade, controlled the movement and choices of labor, created monopolies, and favored some manufacturers and merchants over others. About such regulations and controls Smith says: "The policy of Europe, by obstructing the free circulation of labour and stock both from employment to employment, and from place to place, occasions in some cases a very inconvenient inequality in the whole of the advantages and disadvantages of their different employments."[52] Nevertheless, throughout the *WN* Smith raises questions about a variety of business practices. Smith finds fault with merchants and manufacturers who often act in collusion with one another, with bankers who act with impropriety, and with corporations or "joint-stock" companies whose managers seldom act in the best interests of the owners. He predicts the moral decay of those who make excess profits and finds sloth and indolence among both laborers and those with wealth. At the same time Smith believes that almost no regulation can be advantageous to society,[53] that "the law ought always to trust people with the care of their own interest"[54] and that "though the principles of common prudence do not always govern the conduct of every individual, they always influence that of the majority of every class or order."[55] In these passages in particular, we see Smith describing the faults of business people while at the same time arguing that ideally we should be prudent and left to govern our own economic affairs.

What, then, is Smith's view of an ideal economy? Is Smith a "sunny optimist," as Gunnar Murdal concludes, or does he see a "dark side" to unregulated economic activities? Is the political economy Smith lays out in the *WN* merely a utopian ideal? Did he predict the demise of industrial capitalism, or is he simply a realist about human nature? I shall discuss some of these questions about his analysis of his political economy, the answers to which make a difference to the way in which one reads the *WN*. One need not conclude with Robert Heilbroner that "there is the profound pessimism concealed within Smith's economic and social scheme,"[56] for Smith thought that his political economy could be actualized. He saw that the frailties of human nature are such that no political economy can perfectly achieve its ideal. But he truly believed that whatever the difficulties and inconsistencies of his proposals, they were not insurmountable. A system of economic liberty and natural jurisprudence can provide the best kind of political economy, all things considered, and the alternatives to that kind of society are not promising.

I shall conclude that a careful reading of the *WN* in light of the *TMS* as well as the *LJ* brings to light a rich economic and ethical theory that has often been oversimplified. Rather than being inconsistent, Smith uses these three texts to supplement and collaborate his themes and arguments. I shall argue that the notion of self-interest is complex in the *TMS* and the *WN*, and because the social passions and interests play an equal motivating role, self-interest is not the single dominating motivation in economic affairs. Justice, not benevolence, is the basic virtue in all three texts and the condition for any viable society. The invisible hand, about which so much has been written, is the outcome of market activities, not the "ultimate governor" in economic affairs. Although Smith recognizes problems arising in a free-trade economy, he nevertheless finds such a system to be more propitious to justice, economic liberty, and well-being than other alternatives are. Such an analysis of the *WN* brings into question the philosophical grounds for defending the thesis that an efficient market driven by self-interests or non-tuists can be, in itself, a regulator of economic behavior. No market can function without the framework of justice, and how one acts in the marketplace creates and shapes the character of the very market that allegedly regulates economic behavior.

Smith, then, offers an economic paradigm, but it is not a paradigm often attributed to his thought. He specifically brings into question many of the coveted theses of that kind of position by arguing that we are not motivated merely by self-interests—whether defined as selfishness or merely as an unconcern for others—that we are neither "social atoms" nor merely utilitarians, and that the perfectly competitive market is the ideal result of moral behavior and the constraints of justice but not itself the ideal of moral anarchy. This conclusion demands a thoughtful reconsideration of some of the defenses of various versions of the self-interest views of the *WN*. At the least, one cannot depend on the *WN* as a justification for what C. B. Macpherson has called "possessive individualism," and one would do well to reread what Smith said about motivation, utility, and the invisible hand.

# 1

# The Moral Psychology of the *Theory of Moral Sentiments*

> Morality and therefore moral science exist upon the tension between what the individual may do in his own interest and what he must concede to the interest, or rights, of others as such. In this, as in many fundamental respects, Smith is intelligible as the disciple of Hobbes, the translator of Hobbeanism into an order of society.[1]

This commentary on the *TMS*, written by Joseph Cropsey, sets the stage for the analysis of Smith's moral psychology, for it raises a central issue. How does one read the *TMS*, and how does one then compare the *TMS* with the *WN*? Cropsey solves the problem with the controversial conclusion that Smith is a "disciple of Hobbes" in both works. To deal with this issue, in this chapter I shall examine some of the central concepts in the *TMS*, specifically Smith's notions of self-interest, sympathy, the ideal of the impartial spectator, and the role of the virtues. While acknowledging that the selfish passions are one of the forces of motivation according to Smith in the *TMS*, I shall challenge Cropsey's conclusion that Smith is a psychological or rational egoist. I shall argue that Smith's distinctive contribution to moral psychology is his unique challenge to egoism. He does not set up a dichotomy between egoism and benevolence. Rather, he criticizes both moral theories that derive the basis for moral judgments from self-interest and moral theories that find the sole end of morality to be benevolence. According to Smith, human beings are not motivated merely by selfish passions or self-interest; both prudence (the virtue of self-interest) and benevolence are virtues; and the basic virtue is not benevolence but justice. As we shall see in Chapter 3 the *TMS* lays the groundwork for Smith's development of a political economy, a political economy not merely driven by self-interest in which prudence, cooperation, and an institutional framework of justice provides the necessary conditions for a viable economy.

## Egoism

The term *egoism* often refers to a theory of human nature, a theory that describes, in fact, how we are motivated and how we behave. An egoist argues that human beings are motivated by, and behave in accordance with, their selfish passions, their self-interests, or their own pleasure. The most widely known form of egoism is psychological egoism. The psychological egoist states that in fact most of the time we desire our own pleasure and are motivated by selfish passions or self-interests. The psychological egoist is not merely noting the vague tautology that all one's passions, pleasures, and interests are one's own but, rather, is proclaiming some formulation of the idea that "the ultimate aim of every human action is to obtain some personal benefit for the agent"[2] or, at least, to avoid harm to oneself. Sometimes egoism refers to "ethical egoism," a theory concerning how one ought to be motivated and behave. The ethical egoist asserts that one ought to be motivated by, and act according to, one's self-interests, whether or not in fact one is so propelled.

Egoism is often contrasted with altruism. The psychological altruist claims that in fact one has non–ego-oriented desires and interests and that sometimes one is genuinely motivated to help others whether or not there is a benefit for oneself. The ethical altruist believes that one ought to act benevolently whether or not one is so motivated. Some altruists, such as Francis Hutcheson, contrast benevolence with egoism, arguing that although benevolence is a virtue, acting in one's self-interest is not.

Both Hobbes and Smith have been wrongly interpreted as psychological egoists. As a number of recent commentators have carefully argued, Hobbes does not claim that we are always motivated by selfish passions and selfish interests, that he notices that we have altruistic motivations as well. Indeed, Hobbes recognizes that at certain crucial moments when safety, survival, or self-preservation is at stake—for example, in a state of nature—it would be irrational not to be self-interested. Hobbes, then, is best called a "predominant egoist"[3] or a "rational egoist."[4] Interestingly, in the *TMS*, Smith reads Hobbes as a psychological egoist and even sometimes as an ethical egoist, and then takes Hobbes to task for his views.[5] Cropsey finds Smith to be a Hobbesian throughout all his works, and Smith is often taken to be a psychological egoist or even an ethical egoist in the *WN* and an altruist in the *TMS* by commentators such as Glenn Morrow. As we shall find, Smith amasses arguments against both egoism and altruism, claiming that neither serves as a complete description of moral psychology nor does either by itself serve as grounds for a viable ethical theory.

## Self-Interest, the Social Passions, and Benevolence

Smith beings the *TMS* with the following statement:

> How selfish soever man may be supposed, there are evidently some principles in his nature, which interest him in the fortune of others, and render their happiness necessary to him, though he derives nothing from it except the pleasure of seeing it. . . . The greatest ruffian, the most hardened violator of the laws of society, is not altogether without it.[6]

The "principles of [our] nature" are imagination and sympathy, the latter being derived from what Smith calls the social passions, but sympathy is neither a passion nor the sentiment of benevolence. From the beginning of the *TMS*, then, he explicitly rejects the thesis that human beings are merely selfish and not genuinely interested in others. This "interest in the fortune of others," however, is not identified merely with benevolence, because Smith repeatedly states that

> every man is, no doubt, by nature, first and principally recommended to his own care; and as he is fitter to take care of himself than of any other person, it is fit and right that it should be so. Every man, therefore, is much more deeply interested in whatever

> immediately concerns himself, than in what concerns any other man.[7]

Therefore self-interest is a motivating principle of human action, but it is not the only principle nor is it equated with selfishness. Moreover, although I have genuine interests in others, these may not be entirely benevolent interests. And, as we shall see, self-interested motivations are not in themselves necessarily evil, and benevolence is not the only virtue. But what does Smith mean by self-interest? How does he reconcile the motivation of self-interest with the virtue of benevolence?

Like his predecessors Hobbes and Hume, Smith believed that we are motivated by passions and sentiments, not by reason. But following the seventeenth-century philosopher Anthony Ashley Cooper (Lord Shaftesbury), Smith differs from Hobbes and Hume in claiming that human beings experience three sorts of passions: social passions such as generosity, compassion and esteem; unsocial passions such as hate and envy; and selfish passions such as grief and joy that are connected with our own pleasure and pain. These three sets of passions, the selfish, the unsocial, and the social passions, are the bases for our desires and are driven by pleasure and the avoidance of pain, but later we shall see how Smith qualifies what appears to be a utilitarian moral point of view, first, by sorting out the passions from the interests and, second, by arguing that utility is only one source of moral approbation. Even though all passions are driven by pleasure and the avoidance of pain, often our interests in others result in acts of benevolence, for example, those that give us pleasure for their own sake simply because of our social passions.[8]

One of our social passions is the source of our "desire for approval" by other people:

> Nature, when she formed man for society, endowed him with an original desire to please, and an original aversion to offend his brethren. She taught him to feel pleasure in their favourable, and pain in their unfavourable regard. She rendered their approbation most flattering and most agreeable to him for its own sake; and their disapprobation most mortifying and most offensive.[9]

Interestingly, Smith adds a normative dimension to our natural desire for approval: "Nature, accordingly, has endowed him [a human being], not only with a desire of being approved of, but with a desire of being what ought to be approved of; or of being what he himself approves of in other men."[10] Notice here the emphasis on "nature" and the "natural order." Counter to Mandeville and Hobbes, Smith persuasively postulates that human beings are by nature motivated by selfish, unsocial, and social passions, the last being the source of both the desire to be approved and the desire to emulate what is held to be the standard or ideal. Smith is critical of Hobbes because, he contends,

> according to Mr. Hobbes, and many of his followers [including Mandeville], man is driven to take refuge in society, not by any natural love which he bears to his own kind, but because without the assistance of others he is incapable of subsisting with ease or safety. Society, upon this account, becomes necessary to him, and whatever tends to its support and welfare, he considers as having a remote tendency to his own interest; and on the contrary, whatever is likely to disturb or destroy it, he regards as in some measure hurtful or pernicious to himself.[11]

What is crucial is Smith's point that we are as naturally a social being as a selfish one and that we cannot derive one set of passions from the other.

In reaction to Hobbes, Joseph Butler, an earlier English contemporary of Smith, argued that even if self-interest is the strongest motivating force for human action, it can be neither divorced from a social context nor identified with selfishness. Smith, however, goes further. Remembering the influence of Locke and Hume on Smith, one must be careful not to exaggerate the role of society and social institutions in Smith's philosophy. Nevertheless, it is an error to say that for Smith self-interest is pure unrestrained asocial selfishness or even the single motivating passion or interest. To ignore the role of the social passions in what Smith calls the natural order simply is not in keeping with the arguments in the *TMS*. Although in an obvious way he recognizes that each of us is motivated by self-interest in the sense that all our passions and interests are our own, each of us has social as well as egoistic motivations. Moreover, the social passions are distinct and cannot be derived merely from "a form of other-directed selfishness."[12] I shall explain in Chapter 4 that Smith gives primary importance to the individual rather than the social order, and so it is not incorrect to label him as a moderate individualist. One of the natural qualities of human beings however, is their desire for approval by others and their desire "of being what ought to be approved of." This qualifies the individualist perspective attributed to Smith. The "natural order," then, does not consist merely of a group of autonomous independent selfish individuals. We are naturally part of a social order, and we are genuinely interested in others.

Because the social passions have as much weight as do the selfish ones, it is likely that by self-interest Smith has in mind an eighteenth-century notion of self-love that he, like Butler, distinguishes from selfishness. Like Butler, too, Smith distinguishes various kinds of self-interest.[13] In talking about self-interest, Butler makes several distinctions that help clarify Smith's more obtuse analysis. Accepting, for the sake of the argument, the fact that we are motivated by our self- interest, Butler points out that there are at least four kinds of self-interested motivating factors. First, we often act merely from some unpremeditated passion directed toward an external object. Second, Butler states, we are motivated by self-love. That is, we act with the intention to benefit or avoid pain for ourselves in disregard of others. He points out, however, that not all self-interested actions are selfish, not all selfish actions are beneficial to ourselves, and some selfish actions actually benefit others despite our motivations to the contrary. Third, some of us are genuinely interested in others and therefore are motivated by benevolence. Fourth, we are sometimes interested in following our conscience, that is, doing our duty. (Later in the chapter we shall see how Smith links conscience to the passions.) Butler also separates motivation from moral judgment: When making moral judgments, we evaluate the results of actions, that is, whether an action benefited or harmed ourselves or benefited or harmed others. A good action, according to Butler, however motivated, is one that benefits or does not harm ourselves or that benefits or does not harm others.[14]

Unlike Butler, Smith does not carefully distinguish passions from interests, and thus he has been accused of equating them. According to A. O. Hirschman, this allows Smith to identify selfish passions with self-interest in the *WN*, thereby transforming the latter into the source of public benefits.[15] We shall see that Smith does not confuse passions with interests, although my arguments will be somewhat in the form of conjecture (as are Hirschman's), because Smith does not address this issue directly in Hirschman's terms.

In introducing three sets of passions, Smith borrows from what Shaftesbury called "natural affections." Smith is arguing that "by Nature" human beings are endowed with three kinds of drives. Although these drives or passions are unpremeditated and irrational, as Butler suggests, they are of three sorts, depending on the direction of the passion or the natural affection: to the self, to others, and in negative reaction to others. In this regard Smith writes:

> Revenge [an unsocial passion], therefore, the excess of resentment, appears to be the most detestable of all the passions. . . . [At the same time] man, therefore be naturally endowed with a desire of the welfare and preservation of society [the social passions]. . . . [S]elf-preservation, and the propagation of the species, are the great ends which Nature seems to have proposed in the formation of all animals. Mankind are endowed with a desire of those ends, and an aversion to the contrary [the selfish passions].[16]

Moreover, Smith implies that interests are derived from, but not identical to, the passions. Self-interest or self-love is derived from the selfish passions, but as I shall argue, self-love is not identified with selfishness, because self-love, like the other interests, can be virtuous (the virtue of prudence) or evil (greed or avarice). Our interests in others, including our desire for approval, for what ought to be approved of, and sympathy, derive from the social passions. Again, these interests can be virtuous (the virtue of benevolence) or evil (harm to others). In addition, Smith distinguishes an impartial virtue of the social passions and interests, the virtue of justice. The unsocial passions translate into reactive interests against others. Often these unsocial interests are not a result of what another has done to harm someone but, rather, an aversive reaction to actions that have nothing to do with that person's self-interests. Thus the unsocial interests are distinct from self-interests. Smith would not argue that the unsocial interests can be virtuous except as magnanimity, the self-restraint or self-command of those interests, and their vices exhibit themselves as revenge, vengeance, and the like.[17]

Smith sorts out the passions from the interests in a second way. He uses the term *self-interest* to refer to self-love, to identify selfish interests, or simply to designate those interests with which we are most intimately concerned. Although our passions are more or less equally directed to ourselves and to others, our interests—both our self-interest and interests in others and accordingly their virtues and vices—are circumscribed by social circumstances, by habits, by those closest to us—for example, family, friends, and associates—and by the human limits of rationality and foresight.[18] So social and selfish interests are more proximately focused than are passions. In the *WN* Smith makes a further distinction between private interests and public interests. Private interests are the interests of individuals or groups of individuals, whereas public interests are those interests that concern society or public welfare.

Turning specifically to the question of whether self-love or self-interest is always selfish and arguing against Mandeville and Hutcheson, Smith points out that acting in one's own self-interest can be virtuous. Smith, like Butler, recognizes that not all self-interested actions are virtuous or beneficial to oneself and that sometimes even selfish actions may produce beneficial consequences for oneself or others. But self-preservation, economy, prudence, and other self-interested principles of action are qualities to be admired and emulated, just as benevolence is. "The habits of oeconomy, industry, discretion, attention, and application of thought, are generally supposed to be cultivated from self-interested motives, and at the same time are apprehended to be very praiseworthy qualities, which deserve the esteem and approbation of every body."[19]

When making moral judgments, Smith asserts, motives, principles, and actions are evaluated on the basis of the virtues of self-interest as well as on the principle of benevolence:

> Regard to our own private happiness and interest, too, appear upon many occasions very laudable principles of action. . . . The mixture of a selfish motive, it is true, seems often to sully the beauty of those actions which ought to arise from a benevolent affection. The cause of this, however, is not that self-love can never be the motive of a virtuous action, but that the benevolent

> principle appears in the particular case to want its due degree of strength, and to be altogether unsuitable to its object. . . . Carelessness and want of oeconomy are universally disapproved of, not however as proceeding from a want of benevolence, but from a want of the proper attention to the objects of self-interest.[20]

Because Smith traces the source of benevolence to the social passions, benevolence is independent of self-interest. Smith says in this regard: "How can that be regarded as a selfish passion, which does not arise even from the imagination of any thing that has befallen, or that relates to myself, in my own proper person and character, but which is entirely occupied about what relates to you?"[21] One is not always benevolent merely because it is necessary for one's own well-being but, rather, because one is genuinely interested in the well-being of another. Thus Smith follows Shaftesbury in believing that the social passions are equally motivating and that they are the source of their own interests. So in the important sense it is not true that one always acts in one's own self-interests.

In summary, the strengths of Smith's analysis are that (1) human beings are equally motivated by the selfish, unsocial, and social passions; (2) one's interests, unlike one's passions, are not equally directed outward but tend to focus on the circle of interests in physical, geographical, and social proximity; (3) self-interested motives and actions are evil only insofar as and when they intend or produce harm either for oneself and others; (4) such actions as produce good for oneself as well as those that produce good for others are morally valuable; (5) motive, action, and consequences ought to be evaluated separately, so, for example, a selfish motive that produces social welfare is thereby a good action; and (6) most human actions are "mixed," and one evaluates such actions accordingly. So it could be the case that the best intentions produce negative consequences or, conversely, that evil intentions actually produce some personal or social good.

It is evident, then, that Smith does not equate self-interest with selfishness. The motives of self-interest are complex, and we can be motivated by social passions and interests as well as by our own pleasure. Even the most egoistic person cannot wholly ignore what is approved or desirable by the standards of others. Self-love can be virtuous and produce virtuous actions, both for oneself and for others, and so virtue cannot be identified solely with benevolence or vice with self-love. It is an error, then, to call Smith a psychological egoist, to give a neo-Hobbesian interpretation to Smith's concept of self-interest in the *TMS*, to argue that self-love is inherently bad, or to conclude that Smith derives benevolence from self-interest. Moreover, as we shall elaborate in discussing virtues, it is equally erroneous to claim that Smith extols only the virtue of benevolence in the *TMS*, as the virtue of prudence, not benevolence, restrains the selfish passions and selfish interests and as justice, not benevolence, is the basic virtue.

## Sympathy, the Imagination, and Moral Judgments

In the Introduction we discussed the nineteenth-century "Adam Smith Problem" and some of its more recent counterparts. A misreading of Smith that has contributed to a continuation of this alleged "problem" is the contemporary misidentification of benevolence with sympathy in Smith's moral system. Milton Friedman, for example, contends that "Smith regarded sympathy as a human characteristic, but one that was itself rare and required to be economised. . . . [T]he invisible hand was far more effective than the visible hand of government in mobilising not only material resources for immediate self-seeking ends but also sympathy for unselfish charitable ends."[22] Robert Lamb points out that according to Smith, "the two universal human sentiments experienced by all men"[23] are sympathy and self-interest. And though recognizing that sympathy is not benevolence, Cropsey sometimes equates sympathy with the sentiment of empathy.[24]

Smith's use of the word *sympathy* is indeed confusing, and in the *TMS* he employs the term in at least five ways. That is, *sympathy* is sometimes used in the traditional sense to mean (1) pity or compassion and therefore a source of benevolence, and/or sometimes (2) sympathy is synonymous with empathy. But he wants to break with these traditional usages. Recall that in the first chapter of the *TMS* Smith observes that "how selfish soever man may be supposed, there are evidently some principles in his nature, which interest him in the fortune of others." One of these "principles" is sympathy. At the same time, Smith says, "as we have no immediate experience of what other men feel, we can form no idea of the manner in which they are affected, but by conceiving what we ourselves should feel in the like situation."[25] Because we cannot feel what another is feeling or share that person's experiences, sympathy, in Smith's technical use of the term, is neither a sentiment nor a passion, but rather, an agreement or understanding of sentiments, in Smith's words, a "fellow-feeling with any passion whatever"[26] of another person, only one of which might arouse pity or compassion. Because it is the fellow feeling of any passion whatsoever, sympathy is not identified with empathy or any other passion or sentiment. Sympathy is distinct from self-love because it is not a passion or derived from the selfish passions. Nor is it to be equated with benevolence.[27] Not only is sympathy neither a passion nor a sentiment; sympathy also is the understanding of what another feels: "Sympathy can give rise to selfish as well as unselfish forms of behavior."[28] That is, one can comprehend the selfish feelings of another, approve of them, and act accordingly. In its technical sense, then, sympathy is (3) the recognition of the passions and sentiments of another; (4) "sympathy" with those passions, that is, an agreement with them; and/or (5) approval of that set of passions as an appropriate reaction to a particular situation or kind of situation.

Smith carefully separates himself from his reading of Hobbes on the notion of sympathy. Sympathy does not have as its source, or is it a result of, self-love. "Sympathy . . . cannot, in any sense, be regarded as a selfish principle."[29] When we sympathize we place ourselves in another's situation, not because of how that situation might affect us, but, rather, as if we were that person in that situation. We truly project ourselves into another's experience, according to Smith, in order to understand—although not experience—what another person is feeling rather than merely to relate that situation to our own. Sympathy is the comprehension of what another feels or might feel in a situation, but it is not an experiential or sentimental identification with that feeling. With the help of imagination, sympathy is the means through which we place ourselves in another's situation and conceive of what another feels, experiences, or is capable of feeling in a particular situation or set of circumstances. "Sympathy, therefore, does not arise so much from the view of the passion, as from that of the situation which excites it."[30] Smith is not saying that we do not empathize, pity, or feel compassion for others. In some sense we do share feelings, and sympathy is the source of sharing those feelings, as it allows us to understand the passions and sentiments of another. But the "sympathetic passions" that result from fellow feeling are secondary passions that we experience when we react to what another is feeling, as we cannot actually experience that person's passions or sentiments. Sympathy, then, allows us to understand others' feelings and passions that we cannot experience. Thus sympathy is linked to the social passions, because the latter arouse our interest in others, and sympathy enables us to understand their points of view.

The role of imagination is crucial to understanding Smith's notion of sympathy and indeed his whole moral psychology. Unlike his teacher, Francis Hutcheson, Smith wants to avoid the claim that human beings have a "moral sense" a special psychological faculty that enables us to be sympathetic and benevolent and to make moral judgments. To avoid a moral sense theory Smith argues that each of us has an active imagination through which we can recreate feelings, passions, and the point of view of another. In this imaginative process we do not literally feel the passion of another, but we are able to "put ourselves in another's shoes," so to speak, and to understand what another is experiencing from his or her perspective. The imagination also plays an important role in Smith's theory of the impartial spectator. Whether or not imagination and sympathy are faculties—and so have the metaphysical status similar to the faculty of moral sense—is unanswered in the *TMS*. The important difference between sympathy and the faculty of moral sense is that even if sympathy is a kind of faculty, sympathy is a more general faculty that enables approbation and thus moral approbation, but is not to be identified with either.

Imagination plays another important role in Smith's moral psychology. In the beginning of the *TMS* Smith sometimes talks as if sympathy were a general principle of "fellow understanding," that sympathy would enable us to understand the passions and interests of another even if we felt resentment or even abhorrence toward those passions or toward that person. Imagination is important to this scheme, perhaps more important than Smith admits, because it allows us to project ourselves and understand what another is feeling even when we are revolted by that feeling, that is, not in agreement with it.

In Sections II and III of Part I of the *TMS* Smith appears to equate sympathy with mutual sympathy, with those passions of another with which we share a fellow feeling. Smith recognizes what he calls "the Pleasure of mutual Sympathy,"[31] that is, the pleasure or pain we receive from perceiving similar sentiments in others. This is a source of approbation. Smith differs from Hume in this regard, because he argues that we could sympathize with any kind of passion of another if we felt that the sentiment expressed is one we would feel in that particular situation. Smith writes: "In the sentiment of approbation there are two things to be taken notice of; first, the sympathetic passion of the spectator; and, secondly, the emotion which arises from his observing the perfect coincidence between this sympathetic passion in himself, and the original passion in the person principally concerned."[32] We might sympathetically and vicariously share a passion, for example, anger, in a situation, although we might not approve of its appropriateness. When we both share a fellow feeling and agree with its appropriateness to a particular situation or kind of situation, we approve of that sentiment. Thus sympathy is the source of approval, as we approve of those sentiments with which we agree and agree are appropriate to a given situation. Elsewhere in the *TMS*, however, Smith implies that not all sympathy is mutual sympathy, because we can imaginatively sympathize with any passion of another even if we detest it. Thus sympathy is not identified with approbation because it is more emotionally neutral and can approach or understand all types of feelings and sentiments.

Because sympathy is the primary source of approval or disapproval and because we approve of those passions and sentiments of another with which we mutually agree and disapprove of those with which we disagree, sympathy is one of the four sources of moral approbation or moral judgment. But moral sentiment is not merely determined by sympathy, although without sympathy, moral approbation will not occur. A moral judgment or moral approbation is based on moral sentiment (rather than reason), according to Smith, but reason plays a role. That is, a moral judgment is an informed (rational) judgment based on approbation. Smith argues:

> When we approve of any character or action, the sentiments which we feel, are . . . derived from four sources, which are in some respects different from one another. First, we sympathize with the motives of the agent; secondly, we enter into the gratitude of those who receive the benefit of his actions; thirdly, we observe that his conduct has been agreeable to the general rules by which those two sympathies generally act; and, last of all, when we consider such actions as making a part of a system of behaviour which tends to promote the happiness either of the individual or of the society, they appear to derive a beauty from this utility.[33]

Moral judgments, then, are made first on the basis of a fellow feeling with the motive and second on the basis of mutual sympathy with whoever benefits from the action.

"General rules," the third source of moral judgments, are normative or moral rules, rules that arise from individual mutual sympathies. "The general rule . . . is formed, by finding from experience, that all actions of a certain kind, or circumstanced in a certain manner, are approved or disapproved of."[34] What is normally approved in a society and what that society thinks ought to be approved becomes, by induction, a general rule for that society. A series of mutual sympathies creates a moral rule, and because we desire the approval of others, a moral rule can also, in turn, influence our sympathies and even create a sense of duty or conscience. We obey moral rules because they are both derived from and are the source of approbation.

It is tempting to say that Smith's notion of a general rule is utilitarian, that is, that we adopt and approve of those rules that "work for society" or are in some way beneficial, so that we should approve of and follow those rules that are most useful. Or perhaps Smith is claiming that following a general rule is good for society.[35] On the latter point it is surely true that in the *Lectures on Jurisprudence* he states that peace and order are usually of greater benefit to society than is revolution. It is only in extraordinarily horrific circumstances that following the general rules of a society is not the most prudent course of action, according to Smith, because he finds that in general, revolution is usually to the benefit of only the few who gain power.[36] He is making the negative point that not following general rules—if this entails challenging conventions and government—is usually not in the best interests of society.

On the former point, however—that general rules, being developed out of what is approved or thought to be desirable, always work and/or are always useful—Smith would take exception, for two reasons. First, not all of what we approve or find desirable is useful in any sense. Smith is careful not to identify moral desirability with utility. (This point will be developed in a later section of this chapter.) Second, given Smith's analysis, it could be that a particular society developed a set of general rules that would be harmful to that society. Smith would find this unlikely, as it was his belief that there is an underlying universal natural order of which human beings are a part, but the possibility is not ruled out.

Because we seek the approval of others and "desire of being what ought to be approved of," moral rules—as the general principles arising from what society thinks ought to be approved—serve as the basis for our feeling that sometimes we have duties or obligations, that is, our sense of conscience. Conscience is further buttressed by the idea of the impartial spectator, a crucial concept in Smith's moral psychology that we shall discuss next. The fourth source of moral judgment is utility, but it is only one of the sources of moral approbation. The reason is that the principle of sympathy and the imagination free each of us from the pleasure–pain directives of the passions. Sympathy allows us to separate approval on the basis of motive or "agreeableness" to general rules from approval based on utility.

## The Impartial Spectator and Conscience

The notions of sympathy, approbation, and moral judgment are connected with another important concept in Smith's moral psychology, namely, the "impartial spectator," a notion that places morality firmly in a social context. Smith's impartial spectator is not to be identified with either Hutcheson's "ideal spectator" or Hume's "disinterested benevolent spectator," because Smith's spectator, although virtuous, is neither ideal nor necessarily benevolent. Smith's point is simpler than that of Hutcheson or Hume. He wants to create an explanatory mechanism for the possibility of moral judgments that does not depend on a moral sense theory and that is based on common sense. According to Smith, when making a moral judgment we must be able to disengage ourselves from the situation and from our own sentiments; otherwise the judgment is merely an expression of our subjective feelings of approval. Just as we can understand fellow feelings, so too, Smith observes, our imagination plus the social passions and our natural desire for "what ought to be approved of" provide the conditions to enable us disinterestedly or impartially to approve or disapprove of a motive, character, or action. Sympathy, backed by the imagination and assisted by reason, allows us to disengage ourselves from our own feelings and to evaluate dispassionately both the motives and actions of others and our own motives and actions as well.

Although Smith personifies the spectator, because his spectator functions only in the context of making moral judgments, it is perhaps most appropriate to think of the impartial spectator not as a full-fledged person but, rather, as an abstraction that describes those characteristics that enable us to make disinterested moral judgments. This spectator is that aspect of a well-informed person that enables that person to separate himself or herself from a situation in order to evaluate it dispassionately. The impartial spectator has the foresight to measure this situation in the context of other similar situations, in terms of what society feels it ought to approve and in accordance with its moral rules. The impartial spectator represents the average moral evaluator rather than the ideal and impartial point of view, according to Smith, because moral sentiments out of which the spectator makes his or her judgments arise from shared sympathies rather than from ideal principles. By establishing standards for what ought to be approved, the impartial spectator is also the source as well as the user of moral rules and therefore a source of evaluation. Moral rules are just those principles of moral action generally agreed upon as those that would be acceptable to an impartial spectator, that is, those rules that should be acceptable to society, or what a society feels it ought to approve of. Moreover, according to Smith, benevolence is only one of the criteria for evaluating the motive or actions of another, because prudence and economy, the virtues of self-love, and justice are also important criteria. In summary, in moral decision making, Smith's impartial spectator asks (1) whether a particular judgment would be the kind of judgment that society would deem appropriate to that kind of situation, (2) whether it is the kind of decision that others should make, all things considered, and (3) whether that decision embodies a rule that could serve as a general rule for other similar cases.

The notion of sympathy and the concept of the impartial spectator allow Smith to separate self-interested motivation and the objects of the passions (pleasure and the avoidance of pain) from moral action and moral judgment. This distinction enables him to explain how it is that we often act immorally or irrationally, driven by selfishness or the desire for pleasure, how it is that we often act without selfish or pleasurable recompense, because of the desire not merely to be approved by others but also to be worthy of that approval, and also how it is that we can evaluate motives, actions, and their consequences from a disinterested perspective.

Gilbert Harman argues, correctly I believe, that Smith's notion of an impartial spectator helps explain moral motivation. We are motivated not merely to receive the approval of others. We also have a natural passion for "praise-worthiness [that] is by no means derived altogether from the love of praise."[37] This passion is reinforced by imaginatively seeing ourselves through the eyes of others.[38] Thus the impartial spectator provides the social sanctions for morality, which in turn are the sources of conscience, because the impartial spectator plays an important role in self-evaluation as well as in judging others. Smith explains that we can judge ourselves impartially just as we can impartially judge others. Our natural desire for approval motivates us to self-evaluation, and imaginative sympathy allows us to "step aside" and become impartial self-spectators.

> The principle by which we naturally either approve or disapprove of our own conduct, seems to be altogether the same with that by which we exercise the like judgments concerning the conduct of other people. We either approve or disapprove of the conduct of another man according as we feel that, when we bring his case home to ourselves, we either can or cannot entirely sympathize with the sentiments and motives which directed it. And, in the same manner, we either approve or disapprove of our own conduct, according as we feel that, when we place ourselves in the situation of another man, and view it, as it were, with his eyes and from his station, we either can or cannot entirely enter into and sympathize with the sentiments and motives which influenced it.[39]

The internal impartial spectator is our conscience, what Smith calls "the man within." He identifies following our conscience with following our sense of duty. The authority of conscience, what our duties are, is specified by the general rules of a society. Conscience originates with sympathy and the desire for approval by others and becomes autonomous and independent of our self-interests and thus the source of self-regulation and our sense of duty. Conscience has its source in social relationships, according to Smith, but not in benevolence.

> It is not the love of our neighbour, it is not the love of mankind, which prompts us upon many occasions to the practice of those divine virtues [those prescribed by the impartial spectator]. It is a stronger love . . . the love of what is honourable and noble, or the grandeur, and dignity, and superiority of our own characters.[40]

Because we "desire of being what ought to be approved of," we naturally seek to be honorable and praiseworthy. We value ourselves and value the way others view us, and so we are motivated to follow the rules of the internal impartial spectator, the content of conscience. Thus conscience originates in the selfish and social passions and at the same time controls them. Moreover, as Harman points out, this internalizing of the impartial spectator allows Smith to develop a theory of agent-centered moral motivation linking moral motivation with both our passions and less partial judgments of the spectator without either defining all moral motivation solely emotively or merely dispassionately.[41]

The selfish and social passions, the desire for approval, and sympathy or fellow understanding, then, are the sources of moral sentiments, moral rules, moral judgments, the impartial spectator, and conscience or self-regulation. These concepts permit Smith to make a strong case against egoism as the sole source of morality, to place morality in a social context, and to explain why we are sometimes motivated by a sense of duty and why we can make disinterested moral judgments even about our own actions. Moreover, as we shall see, the notions of sympathy and the disinterested spectator allow Smith to argue that the principle of utility is one, but not the only one, principle of moral approbation.[42]

One of the questions that arises in reading the *TMS* is whether there is a circularity in Smith's moral theory. The impartial spectator is not an abstract ideal; rather he or she is the ideal average moral actor and judge, and the content of his or her actions and judgments are specified by the particulars of a situation and by previous sets of judgments that have created a society's general rules, moral rules that evolve further through the judgments of the spectator. Even as an impartial spectator one cannot escape one's particular social context in making moral judgments, and this social context provides both content for the moral judgment and a precedent for future judgments. This leads some commentators to argue that according to Smith, human beings create morality, moral judgments, and principles of moral conduct rather than discover "moral facts" or moral truths, that morality is inherently contextual and relative to a particular society and, in this sense, is circular.[43] We shall return to this point when we consider Smith's alleged relativism.

## The Virtues: Prudence, Benevolence, Justice, and Self-Command

Smith revised the *TMS* several times, and in the sixth edition of the *TMS* he added the notion of self-command, a quality without which one cannot be "perfectly virtuous." To understand this concept we need first to explain Smith's theory of virtue. His theory of virtue traces its roots to the Stoics, of whom Smith was a great admirer. The complexity of that relationship, and indeed the role of virtue in Smith's works, particularly in the *WN*, is a theme that I shall not develop here.[44] Rather, I shall concentrate on those aspects of his theory of virtue that illustrate his distinctions among selfishness, self-interest, prudence, benevolence, and justice.

Smith is interested in the sources of virtue, what a society held to be virtuous character, and what constitutes a virtuous person. What is virtuous, he feels, are those qualities admired and esteemed by the impartial spectator in accordance with society's general (moral) rules. Virtue, then, is connected with sympathy, which is a source of the moral approbations of the impartial spectator and a source of general rules. But Smith distinguishes virtue from propriety. Propriety is what is approved of or is appropriate. Moreover, propriety is a necessary condition for an action or person to be virtuous, as inappropriateness would rule out the virtuousness or morality of the action or character. But 'though propriety is an essential ingredient in every virtuous action, it is not always the sole ingredient. . . .[45] There is . . . a considerable difference between virtue and mere propriety; between those qualities and actions which deserve to be admired and celebrated, and those which simply deserve to be approved of."[46] Virtues, then, are those superior qualities that are esteemed by the impartial spectator, what an impartial spectator and society determine as those qualities that ought to be approved of.

Smith states that there are four kinds of virtues, prudence, the virtue of the selfish passions and self-love, two virtues of the social passions and interests—benevolence and justice—and a fourth special sort of virtue, self-command. For Smith, "every man, as the Stoics used to say, is first and principally recommended to his own care, and every man is certainly, in every respect, fitter and abler to take care of himself than of any other person."[47] It is as important, morally important, to look after the well-being of oneself as it is to care for others, and so prudence, the virtue of self-interest, is a virtue to be admired.

Yet because morality pertains to social relationships, virtue, too, is connected with society. Benevolence goes beyond not merely harming another, and according to Smith, it is part of our nature to feel obliged to help others. At one point early in the *TMS*, Smith declares "that to feel much for others and little for ourselves, that to restrain our selfish, and to indulge our benevolent affections, constitutes the perfection of human nature; and can alone produce among mankind that harmony of sentiments and passions in which consists their whole grace and propriety."[48] In this context Smith emphasizes that benevolence is a natural affection, just as selfishness is, because the social passions carry as much weight as do the selfish and unsocial ones. Smith describes the virtue of benevolence as a series of commitments related to one's interests: first to one's family, then to friends and neighbors, to one's country, and finally to the universe. Smith recognizes, too, that benevolence is usually considered the superior virtue. But Smith is critical of any moral philosophy that declares that benevolence is the only virtue, that dismisses self-love as an improper motive, or that fails to recognize the virtue of prudence.[49] And Smith is careful to note that "the administration of the great system of the universe . . . the care of the universal happiness of all rational and sensible beings, is the business of God and not of man."[50] So benevolence as a virtue is itself prudent and self-restrained.

Even though benevolence is an important virtue, Smith believes that justice plays the most central role of the virtues. Justice arises from the social passions, but it is not merely interest in, or desire for the approval of, others. The notion of justice, though not well developed in the *TMS* or in any of Smith's writings, is the virtue of impartial social interests. Justice in that respect is the "consciousness of ill-desert,"[51] what an impartial spectator would deem to be appropriate and fair principles governing social relationships. At least in the *TMS*, justice is an individual virtue, an internalized ideal of behavior, and "the main pillar that upholds the whole edifice [of human society]." In the *LJ* and the *WN* Smith does not talk about justice as a virtue but, rather, as natural jurisprudence reflected imperfectly in laws necessary to secure persons and their rights and properties from harm and to enforce fair contracts. As a virtue and a principle governing social and political relationships, justice both is a negative principle that proscribes harming another and includes the positive notion of what Smith calls "fair play": "Mere justice is, upon most occasions, but a negative virtue, and only hinders us from hurting our neighbour. . . . We may often fulfil all the rules of justice by sitting still and doing nothing."[52] At the same time, according to Smith, if one is just,

> he would act so as that the impartial spectator may enter into the principles of his conduct. . . . In the race for wealth, and honours, and preferments, he may run as hard as he can, and strain every nerve and every muscle, in order to outstrip all his competitors. But if he should justle, or throw down any of them, the indulgence of the spectators is entirely at an end. It is a violation of fair play, which they cannot admit of.[53]

One is allowed, then, to compete, to acquire wealth and property, even to indulge one's self-interests, but only if one does not violate agreed-upon principles of conduct (moral rules) that would allow others to compete and indulge their self-interests equally. Similarly, as I shall elaborate in Chapter 2, the laws of justice are negative laws to prevent harms, to protect perfect rights, and to preserve the precepts of fair play.[54] However, Smith does not develop a concept of distributive justice, and given his ideal of a political economy, this oversight must have been a deliberate one.

Although the other virtues are vague and not enforceable by law, Smith claims that "there is, however, one virtue of which the general rules determine with the greatest exactness every external action which it requires. This virtue is justice. The rules of justice are accurate in the highest degree, and admit of no exceptions or modifications."[55] Thus one can develop specific and complete rules of justice, although Smith failed to complete this task in his own writings. According to Smith, justice is the only virtue for which one may justifiably use external force to secure, because only principles of justice apply equally and impartially to everyone. One need not be benevolent in order to be virtuous, but one has a duty not to harm others and to engage in fair play. That is, "beneficence, therefore, is less essential to the existence of society than justice. Society may subsist, though not in the most comfortable state, without beneficence; but the prevalence of injustice must utterly destroy it."[56] So justice is the foundation of the other virtues, because without justice society could not continue.

Glenn Morrow argues that Smith has a duel theory of virtue: "Virtue he [Smith] regards as twofold, consisting of self-interest regulated by justice, and the higher virtue of benevolence."[57] According to Morrow, Smith contends that justice is a primary virtue, because society cannot exist without it, because justice can be exhorted by force, and because justice can be spelled out precisely with rules and laws. Benevolence is a secondary virtue because it is not required in order for society to function, it cannot be insisted upon, nor can it be instituted in a code of rules or law. Justice, however, is merely a negative virtue, and as Smith admits, benevolence is often the most admired. Therefore Morrow interprets Smith as propounding "the moral superiority of benevolence."[58] Morrow's aim is to show that one of the differences between the *TMS* and the *WN* lies in the conception of virtue, benevolence being the superior virtue in personal relations in the *TMS*, and self-interest governed by justice as controlling the "inferior economic virtues" espoused in the *WN*.

Morrow has correctly captured Smith's juxtaposition of justice and benevolence as primary and secondary virtues. Justice and benevolence are not the only virtues, however, and it is unclear which of them is superior. Smith admits that benevolence is the more admired, but at the same time he praises justice as the foundation of the other virtues. He repeatedly emphasizes the importance of prudence, the virtue of the selfish passions, and the result of a good conscience, and it is prudence, not justice, that controls self-interest, at least in the *TMS*.

The fourth virtue, self-command, a virtue that Smith added to the last edition of the *TMS*, is perhaps the most central. About self-command he says: "Self-command is not only itself a great virtue, but from it all the other virtues seem to derive their principal lustre.[59] . . . The man who acts according to the rules of perfect prudence, of strict justice, and of proper benevolence, may be said to be perfectly virtuous."[60] Self-command combines knowledge, foresight, and, most importantly, self-reliance and self-control. By bringing all one's passions under control, coupled with knowledge and foresight, one "may be said to be perfectly virtuous." The perfectly virtuous person would be the perfect impartial spectator, Smith's ideal average person in all aspects. Self-command is the virtue of "superior prudence," for, Smith observes: "Prudence is, in all these cases [of self-command], combined with many greater and more splendid virtues, with valour, with extensive and strong benevolence, with a sacred regard to the rules of justice. . . . It necessarily supposes the utmost perfection of all the intellectual and of all the moral virtues."[61] Moreover, unlike prudence, justice, and benevolence—often appreciated for their utility or effects—self-command is the only virtue valued solely for its own sake.[62] Yet as Smith himself notices, one could be in self-command without being perfectly virtuous if one were not at the same time also perfectly prudent, benevolent, and just. So self-command is a great virtue, the "luster" of the virtues, and itself distinct.[63]

One cannot conclude, then, that benevolence is the superior virtue in the *TMS*. Smith often talks as if prudence and justice are as important as benevolence is, and indeed, self-command appears to be the most likely candidate. Just as Smith is critical of egoism as either a psychological description or a moral theory, so too he questions any moral theory that finds no virtue in self-love and proclaims that benevolence is the only virtue.[64] Thus Smith is neither an egoist nor an altruist but recognizes the motivations and virtues of both. I shall argue in Chapter 3 that the difference between the *TMS* and the *WN* does not lie in different emphases of the virtues and that self-interest, the social passions, and justice are crucial to both texts. Rather, in the *WN*, benevolence—but not the social passions—is irrelevant to economic affairs, and sympathy and the impartial spectator drop out in any discussion of a political economy.

## The Principle of Utility

Among the many readings of the *WN* is the view that Smith is primarily a utilitarian. Although Jeremy Bentham probably coined the term *utilitarianism*, the concept of utility has its roots in ancient philosophy, and Smith discusses that notion in all his works. In brief, the conclusion that he is a utilitarian derives from his apparent claim in the *WN* that autonomous self-interested economic actions in a free market produce economic prosperity for everyone. Although in a later chapter I shall qualify this interpretation of self-interest and question the role of the market as a nonhuman arbitrator, it would be rash to contend that the *WN* does not present an economic theory that takes, at least in part, a utilitarian perspective. But it is interesting to look at the role of utility in the moral psychology of the *TMS* to see how Smith places it in a restricted perspective, a perspective that at least implicitly carries over to the *WN*.

In the *TMS* Smith argues that utility is one, but only one, of the sources of moral approbation, as we have seen. Moral approval derives first from the motives of the agent with which we are mutually sympathetic—that is, those motives that are similar to our own reactions in a similar situation—second from the benefits of the action, third from coincidence with moral rules, and only lastly from the utility of the action. Smith does not deny that utility pleases but "that is not the view of this utility or hurtfulness which is either the first or principal source of our approbation and disapprobation."[65]

According to Smith, there are at least four reasons that utility is only one of the sources of moral approbation. First, we often admire virtue, appropriateness, propriety, and even the beauty of utility for their own sake rather than for what they produce. For example, we admire the architectural wonder of a building or the fitness of a machine.[66] Second, what is approved by the impartial spectator is not always what is useful, although utility is one of the criteria for approval. When making a moral judgment we approve of the appropriateness of the motive as well as the action or its consequences. "To the intention or affection of the heart, therefore . . . all praise or blame, all approbation or disapprobation, of any kind, which can justly be bestowed upon any action, must ultimately belong."[67] Smith qualifies this remark by admitting that only "the great Judge" can evaluate correctly motives and intentions. "Actions, therefore, which either produce actual evil, or attempt to produce it, are by the Author of nature rendered the only proper and approved objects of human punishment and resentment."[68]

As I pointed out earlier, general or moral rules, those rules that a society adopts as governing what ought to be approved of, are not necessarily utilitarian. Thus a society could adopt a set or subset of rules that were not necessarily in its interests.

Third, some virtues, particularly the virtues of prudence and self-command, are not identified with utility, because even if one could make a strong defense of the claim that every action approved by the impartial spectator has a utilitarian component, virtues are those qualities not merely approved of; that is, they are not merely propitious but are those qualities esteemed or admired, often for their own sake. Interestingly, too, because benevolence is only one of the virtues, Smith conceives of the approbation of utility as applying to both oneself and others. A selfish action would receive moral approval if it were beneficial to oneself even if it benefited no one else. Thus, although utility is a source of moral approbation, it is not the only source or the only basis for moral judgments, and the principle of utility is a source of evaluating egoistic as well as beneficent motives and actions.

Fourth, if moral virtue were not distinct from utility, morality would be a more general notion applying to all areas in which utility has a role in judgment.

> For . . . it seems impossible that the approbation of virtue should be a sentiment of the same kind with that by which we approve of a convenient and well-contrived building; or that we should have no other reason for praising a man than that for which we commend a chest of drawers.[69]

That is, what is useful is neither always an object of moral approval nor what is useful always virtuous. Therefore, according to Smith, the principle of utility is not the primary basis for morality or moral judgments. Later we shall see what happens to this kind of qualified utilitarianism in the *WN*.

## Smith's Methodology and the Natural Order

If we look for the foundations of Smith's moral psychology, we will discover that Smith is less concerned with the metaphysical foundations of his philosophy or its epistemological basis than with showing how we are motivated, interact with one another, make moral judgments, and develop a conscience. Although Smith claims to be an empiricist, the *TMS* clearly illustrates Smith's eclectic methodology. It is said that "the work of Adam Smith is generally regarded as a clear example of the Enlightenment practice of adapting beliefs in natural law, benevolent Providence, and Newtonian mechanics to the study of society."[70] It is evident that Smith was influenced by both Newton and Hume. But unfortunately there is little evidence in either the *TMS* or the *WN* that he was overly concerned about method. In the *WN* he supposedly adopts a Newtonian methodology,[71] but one could scarcely say that he adopts a deterministic or mechanistic view of human nature, for in the *WN* he argues that individuals can be free—that they have or should develop their "natural liberty" to do as they please, restrained only by the precepts of natural justice. Thus Smith's interest in Newtonian methodology does not extend to raising questions about the limits of mechanical determinism.

In a persuasive article, J. Ralph Lindgren argues that Smith is an epistemological conventionalist. According to Lindgren, Smith accepts Hume's radical empiricism and argues that we give order to nature and social relationships through habit, custom, and imagination. For example:

> When two objects have frequently been seen together, the imagination acquires a habit of passing easily from the one to the other. . . . Though, independent of custom, there should be no real beauty in their union, yet when custom has thus connected them together, we feel an impropriety in their separation.[72]

According to Lindgren, however, Smith separates the methodology of the natural sciences from that of the social sciences. The natural sciences disinterestedly order and organize data, whereas the social sciences take into account the propriety of an action, "the life situation of . . . fellow man."[73]

Lindgren's thesis is a fascinating one, and it is textually supported by one of Smith's early writings, the "History of Astronomy." Yet one could hardly say that Smith is a conventionalist, at least in the *TMS*, for he talks about the natural order and Divine Providence. Smith also does not worry about the consequences of Hume's radical empiricism. For example, he does not worry about the nature of the self or the problem of memory or that there might be a difficulty with the status of the individual as a moral agent. In truth, he is neither a metaphysician nor an epistemologist, and those sorts of issues do not overly concern him.

Rather, Smith seems to adopt a neo-Aristotelean naturalism without arguing for or defending its underlying metaphysical assumptions: "It appears that, by genus, nature is an internal principle [for Smith], or an internal impulsion in virtue of which all living . . . things act or move as they do."[74] Although it is difficult for the twentieth-century mind to understand the idea of "nature" or a "natural order," what Smith had in mind is a static notion of human nature and a natural order of society, an order that is both universal and unchanging.

For Smith, each of us, everywhere as human beings, has a similar natural endowment, and each of us is directed toward the same "great ends":

> Self-preservation, and the propagation of the species, are the great ends which Nature seems to have proposed in the formation of all animals. . . . Nature has directed us to the greater part of these by original and immediate instincts . . . without any consideration of their tendency to those beneficent ends which the great Director of nature intended to produce by them.[75]

There is a natural order in the universe, an order underlying human relationships. The universe is governed by a deity, although Smith does not spell out the exact relationship between God and nature. Hume criticized Smith for his pious appeal to a deity in this regard, but a more benign reading is to conclude that Smith finds some universal characteristics in human nature, characteristics out of which moral sentiments and general rules develop. Human beings are part of the natural order and harmony of the static and consistent universe. God, of course, is the architect or perhaps the conductor of the natural order. The methods we devise to achieve these ends vary among persons and among different species, and Smith has an evolutionary view of history and economic and political development, as we shall outline in Chapter 2. Nevertheless, the ends for which we strive, for example, "self-preservation and the propagation of the species," translated into economic well-being in the *WN*, are derived from our "nature."

Following the Stoics, Smith states that "every individual . . . naturally prefers himself to all mankind."[76] Yet at the same time he also declares that "it is thus that man, who can subsist only in society, was fitted by nature to that situation for which he was made."[77] Human beings are by nature self-interested and social beings, and there is a natural order of social relationships. What that natural order is, is unclear, but the natural order of society would appear to be that order closest to human nature—to the way we are inherently, or "by nature." In addition, a "sacred and religious regard not to hurt or disturb in any respect the happiness of our neighbour, even in those cases where no law can properly protect him, constitutes the character of the perfectly innocent and just man."[78] The natural order of society is the harmonious system of the passions, interests, and sympathy based on natural jurisprudence within which individuals are free to pursue their interests so long as they act fairly and do not harm others, here remembering that individuals naturally have both social and selfish passions and interests. In the *Lectures on Jurisprudence* Smith adds natural rights, including liberty, to our inherent characteristics. So in the natural order of society individuals have the "natural liberty" to pursue their interests, restrained only by justice.

The natural order of society is the ideal moral order, and Smith seems to imply that there is an ideal that morality, society, and even the political economy should emulate. This ideal is not to be equated with Locke's or Hobbes's presocial or primitive state of nature or with the organicism of Rousseau or Hegel.[79] Rather, it is the order that ideally reflects most closely the natural order, enabling individuals and society to flourish unrestrained by cumbersome artificial conventions. There is, then, a built-in teleology in human nature, an ideal to which we all do or should strive. It is this natural order that is the basis for, and the end of, human nature and thus morality and a political economy.

## Smith's Relativism

Before turning to the *WN*, we need to discuss briefly the sticky question of whether or not Adam Smith was an ethical relativist. This issue is important to evaluating both the status of the moral psychology of the *TMS* and the scope of the economic philosophy of the *WN*.

Smith appears to be an ethical relativist, because moral approval, the content of moral principles, and general rules evolve out of a series of mutual sympathies, so that what one society holds as its most cherished principles need not coincide with another's.[80] The operations of sympathy are the source of moral approbation, and the actions of the impartial spectator that result in moral rules and the development of moral principles arise from individual approbations in specific societal contexts. Smith argues further that moral approbation depends on social relationships. Human beings in solitude are unable to make moral judgments because they have no "social mirror" through which to seek approval and compare themselves.[81] Even one's conscience is created from the social relationships out of which self-evaluation evolves. Because a society's moral rules are based on what is approved by that society and what is agreed upon as to what should be approved, the moral rules and judgments of the impartial spectator are relative to the particular society in which they are developed. Moral rules are, in that sense, empirically generalizable, though of course, they are not necessarily utilitarian. The impartial spectator, being merely the ideal average person in a particular society, is a product of society.[82] Thus it appears that according to Smith there can be many different kinds of moral rules and many different sorts of impartial spectators in each different society, each distinct and even incommensurable to the other.[83]

On the other hand, the terms *nature* and *natural order* play important roles in Smith's philosophy and bring into question his alleged relativism. We are endowed "by nature" with selfish, unsocial, and social passions, imagination, and sympathy. Smith argues that morality and general rules arise from human nature, from the "harmonious order of nature,"[84] and therefore reflect that order. Because moral rules (what Smith calls "general rules") are derived from these faculties that are universally the same in all human beings and because, according to Smith, we generally agree on what ought to be the case, that is, what should be embodied in a moral rule, we do not merely create morality or general rules. Rather, moral rules reflect universal principles or, in Smith's own words, "the will of the Deity,"[85] and the best set of rules most closely reflects the natural order. In the *TMS* Smith is optimistic that human society can closely emulate the harmonious order of nature. In the *WN* the ideal economic order should also reflect the natural order, but he is more pessimistic and finds that that order often fails to be realized in a political economy.

Given this notion of the natural order as the foundation for morality, it is not irrational to conclude that general rules in some way reflect a general or universal order, although how much and in what sense are not clear. Smith perhaps places too much faith in the unity of human nature rather than in our differences. Thus Smith would be surprised to be called an ethical relativist, although his moral psychology leaves room for the possible development of incommensurable moral frameworks in different societies. Nevertheless, his view of the universality of human nature and the natural order precludes a regressive relativism often attributed to his moral psychology.[86]

## Conclusion

The *TMS* lays the groundwork for Smith's later work, the *WN*. Therefore it is important to see how he develops a number of concepts in the *TMS* that play a role in the *WN*. According to Smith's moral psychology, human beings are motivated by three distinct kinds of passions, none of which is primary. Therefore self-interest is not the strongest or the only motivating force, and Smith cannot be called an egoist. The selfish and social passions, when restrained, are the sources of the virtues of prudence, benevolence, and justice. Justice is the basic virtue, as it is what governs internal well-being and is the ground for any viable society. Sympathy, a notion crucial to the *TMS*, is neither a passion nor a sentiment, nor is it the source of benevolence. Sympathy is a source of approval, and what is approved of—and thus the moral rules develop out of what a particular society judges to be what should be approved of. At the same time the underlying natural order guarantees universal similarity in the moral rules. So Smith is not a relativist.

In summary, some of Smith's significant contributions to moral psychology is his attempt to solve the egoism–altruism debate by means of his denial of psychological egoism as a viable theory of motivation, his argument that natural affections or passions are neither interests nor virtues or vices, his refusal to identify morality with altruism, and his recognition that prudence is the virtue of self-interest. Although Smith's attempt to avoid a moral sense theory may not have been altogether successful, because imagination and sympathy seem to operate as moral senses, his definition of the impartial spectator as neither ideal nor benevolent is an innovative approach to "spectator" theory and helps to resolve the problem of moral motivation that is neither subjectivist nor impersonal. Finally, his claim that justice, not benevolence, is the basic virtue sets up the groundwork for the concept of a political economy he develops in the *WN*.

That Smith is consistent in his use of the notion of self-interest in the *TMS* and the *WN*, that justice is the basic virtue in both works, and that thus there is not an "Adam Smith Problem" will be topics for Chapter 3. In the next chapter we shall examine some important themes Smith develops in the *Lectures on Jurisprudence*, themes that are implicit in, and important to, the *WN*.

# 2

# Natural Rights, Liberty, and Natural Jurisprudence

> The first and chief design of all civill governments, is . . . to preserve justice amongst the members of the state and prevent all incroachments on the individualls in it, from others of the same society.—(That is, to maintain each individual in his perfect rights).[1]

A well-defined notion of natural liberty, the free movement of labor, and the assumption that each of us has rights to private property are essential to the arguments in the *WN* and to the kind of political economy that Smith spells out in that work. Indeed, Smith's political economy would make no sense without these assumptions, and in the *WN* he talks about the importance of natural liberty and the "sacred" rights to labor and property. Yet his theory of rights, important as it is to dominant themes in the *WN*, is explained only in the *Lectures on Jurisprudence.* (*LJ*). Smith seldom mentions rights or natural rights in the *WN*. One may only surmise that he presupposes the theory of rights described in his lectures and that he assumes that his readers are familiar with the prevailing theory of natural rights.

Recently, the importance of the role of justice in the *WN* has received attention in the literature on Smith's work.[2] In the *WN* an institutional system of justice underlies natural liberty, labor, and property rights. Smith says in that regard, "upon the impartial administration of justice depends the liberty of every individual."[3] But his theory of justice, crucial to the development and maintenance of a political economy, is clearly articulated only in the *TMS* and *LJ*. Moreover, although justice is a theme reiterated in all Smith's works, its role in a market economy has not always been given due consideration.

In this chapter I shall discuss how Smith's theory of rights developed primarily in the *LJ*, provides a basis for his oblique but important references to rights in the *WN*. I shall pay particular attention to his idea of natural or perfect liberty and his theory of property rights. Justice or natural jurisprudence is intimately connected with Smith's theory of rights, and as far as possible given the limitations of the text, I shall analyze his notion of justice, relying on the *TMS* as well as the *LJ* to flesh out this understanding. Then in Chapters 3 and 6 I shall elaborate on the roles of liberty and justice in a commercial society and, in particular, in a market economy.

Smith's notion of property raises some questions, and near the end of this chapter I shall consider whether his idea of economic liberty and his definition of property as a perfect right conflict with principles of justice that protect public interests. He says that "the establishment of perfect justice, of perfect liberty, and of perfect equality is the very simple secret which most effectually secures the highest degree of prosperity to all the three classes [laborers, landlords, and manufacturers]."[4] At the same time, he declares that "wherever there is great prosperity, there is great inequality,"[5] so that one of the tasks of civil government is to protect property rights or to provide a "defence of the rich against the poor."[6] If one has rights to acquire and secure property and to pursue freely one's economic ends, in this context how can justice, public interests, the satisfaction of basic needs, and equality be served as well? Or as Istvan Hont and Michael Ignatieff frame the question, "How was extreme inequality of distribution in [Smith's] modern society compatible with the satisfaction of the needs of the poorest working members?"[7] Smith recognizes these difficulties, and we shall see whether he resolves them satisfactorily.

A historical note about the *LJ* is important to weighing the force of Smith's remarks in these lectures. During his lifetime he had hoped to write a treatise on jurisprudence tracing the origins of law, civil government, and property rights and laying out a full-fledged theory of justice. It is often speculated that he meant in his work on jurisprudence to constitute the link between the *TMS* and the *WN*,[8] and placing the *LJ* between the *TMS* and the *WN* makes sense. But his aspiration to complete a work on jurisprudence was never realized, and when Smith died, at his own order all his papers and notes were destroyed. Subsequently, however, two sets of student lecture notes have been discovered that document Smith's lectures of 1762–63 (hereafter referred to as *Lectures on Jurisprudence (A)* or *LJ(A)* and his lectures of 1763–64 (hereafter referred to as *Lectures on Jurisprudence (B)* or *LJ(B)*. Many topics in the two sets of notes overlap, although the (A) set, discovered only in 1958, is more complete. One should be cautious, however, because the *LJ* are lecture notes taken by two—albeit conscientious—students and so should not be thought of as the final statement of Smith's theory of justice. Nevertheless, an analysis of issues treated primarily in the *LJ* is necessary to give the background for a number of concepts.

Smith's theory of rights, as will become evident, is borrowed from his predecessors. In particular, much of what he says about justice, property, and the alleged original or social contract depends on his reading of David Hume.[9] Yet there are significant differences between Smith and Hume on the origins and utility of justice, property rights, the property of one's labor, and reasons for the failure of a social contract theory. These differences, I believe, require a careful consideration of Smith's theory on its own merits as a significant contribution to political thought.

## Natural Rights and Natural Liberty

Smith's theory of natural rights articulated in the *LJ* is not an original thesis, nor does he make such a claim. Indeed, he does little to explain the origin of natural rights because, he says, "the origin of natural rights is quite evident. That a person has a right to have his body free from injury, and his liberty free from infringement unless there be a proper cause, no body doubts. But acquired rights such as property require more explanation."[10] Adopting Pufendorf's and Hutcheson's distinction between natural and adventitious rights, Smith states that natural rights are those rights given to us because of the natural order, whereas adventitious rights are acquired and depend on particular political or institutional circumstances.[11] Recalling the discussion of the natural order in Chapter 1, it seems that natural rights are those we have "by nature," thus deriving their origin from the natural order. An ideal system of natural jurisprudence, then, would be the one most closely emulating the natural order, the one most propitious to the exercise of natural rights and their protection from harms. Adventitious rights are conventional rights that vary from society to society and in different historical periods.

Like Smith's concept of justice in the *TMS* that we described in Chapter 1, Smith's theory of rights is, by and large, what contemporary philosophers call a theory of negative rights. Smith derives rights from duties, and he links rights and duties to principles of jurisprudence for enforcement. He reaches the notion of a right from what he takes to be duties to prevent harm. He argues that the first duty of any government is to preserve justice, that is, to protect its citizens from harms.[12] Rights are just those claims or entitlements whose violation would constitute a harm, injury, or unfairness.

This definition of rights allows Smith to distinguish between perfect and imperfect rights. Perfect rights are those rights that cannot be denied, and so their enforcement is morally required. Smith says that "perfect rights are those which we have a title to demand and if refused to compel an other to perform."[13] All natural rights are perfect rights because their violation requires redress. In contemporary analysis, perfect rights are those rights that entail second-party duties to respect or enforce them. Alternatively, perfect duties are mandatory obligations to protect perfect rights to what is to be protected. For example, according to Smith, the right to personal liberty is a perfect right because one cannot be excused from the duty not to infringe on that right. Imperfect rights, on the other hand, are those rights that one should have, but one is not required to defend them; thus only an imperfect duty is thereby implied. An example of the latter is the right to welfare assistance. According to Smith, it is virtuous to help the poor, but it is not obligatory to do so. As we shall see, he thought that jurisprudence is restricted to commutative justice. Laws of justice should be reserved for the exercise of perfect duties to protect perfect rights. Imperfect rights, on the other hand, pertain to distributive justice, rights that Smith reserves for morality rather than for law or jurisprudence.[14]

Smith distinguishes three ways in which a person may be unjustly injured and, thus, three types of rights. One may be injured as a person, injured as a family member, or injured as a citizen. Correspondingly, there are three kinds of rights: individual rights, rights as a member of a family, and rights of citizenship. Only the rights of persons as individuals qualify as natural rights, because all others "depend upon some human deed or institution,"[15] and even then, not all individual rights are natural rights.

There are three kinds of individual rights: the rights to one's person, to one's reputation, and to one's estate, again according to how they can be violated. As an individual one has the right not to be bodily harmed and the right not to have the "free use of one's person" encroached upon[16]; that is, one has a natural right to personal liberty. These two rights, that is, the rights to one's body and to liberty, along with the right to protection from injury to one's reputation, Smith regards as natural rights. They are given to everyone; they are ahistorical; and there are no exceptions to claims to those rights. Smith admits, however, that these rights can be unrecognized or violated.[17] "Estate" or property rights are adventitious rights, because they do depend on institutions for their definition, scope, and protection. (The Appendix lays out in schematic form Smith's system of rights.)

It is commonly held that liberty is the basic and most important right in the *WN*. I shall argue that justice, not liberty, is basic. Nevertheless, the crucial role of economic freedom cannot be overlooked, and "natural" or "perfect" liberty is a central concept in the *WN*. The natural right to the free use of one's person is spelled out in the *LJ*:

> The right to free commerce, and the right to freedom in marriage, etc. when infringed are all evidently encroachments on the right one has to the free use of his person and in a word to do what he has a mind when it does not prove detrimental to any other person.[18]

In the *WN* natural liberty is central to Smith's idea of a modern commercial economy: "All systems either of preference or of restraint, therefore, being thus completely taken away, the obvious and simple system of natural liberty establishes itself of its own accord."[19] Liberty can be traced to one's natural drives for self-preservation and propagation of the species, coupled with Smith's statement in the *TMS* that "every man is certainly, in every respect, fitter and abler to take care of himself than of any other person." Natural liberty is not identified with self-interest but, rather, with the ability to choose and control all one's interests. Recall that in the *TMS* the natural order is the social order that preserves most closely human nature—how individuals are naturally endowed. In the *TMS* we learned that human beings are endowed with selfish, unsocial, and social passions and interests, reason, a strong imagination, and sympathy. In the *LJ* Smith adds certain rights to these natural endowments. In the *WN* the "obvious and simple system of natural liberty" is the harmonious order of individuals enjoying their freedom to take care of themselves and engage in free commerce. Natural liberty, then, is the realization of one's natural ability to control one's own life.

Smith is not concerned with the epistemological problem of how much natural liberty can operate in a causally ordered Newtonian universe. If one is naturally free, will one be free from one's own desires and passions and the natural and social order in which one finds oneself? Such an anarchical freedom from oneself and from the social order is not what Smith has in mind, yet he did not adopt a pure Newtonian determinism, either. Similarly, when he describes a political economy, he sometimes seems to be presenting an "economic machine based on the norm of natural liberty,"[20] without seeing any contradictions in such a scheme. How liberty as self-control and self-determination functions in the natural order or a Newtonian order is a problem, like many that crop up in the *TMS* and the *WN*, that unfortunately is not resolved in Smith's philosophical system.[21]

In the *WN* Smith often refers to liberty as "perfect liberty," in which perfect liberty is natural liberty in the context of economic and social activities. Perfect liberty exists "in a society where things were left to follow their natural course,"[22] that is, in a market economy.[23] But perfect liberty, when allowed to take its "natural course," should operate only in the context of the restraints of justice, but not restraints on trade, bartering, manufacture, or wages, rather, restraints at the more basic level of security from harm and unfair play. The reason is that freedom is neither a virtue nor a vice. Freedom is the absence of external constraint, but how one exercises that independence, not the independence itself, determines the virtuousness (or evil) of an action. Therefore freedom cannot be equated with virtue or excellence, but neither can it be thought of as "a discharge from the inhibitions that traditionally were known as virtues."[24]

According to Smith, natural liberty as economic freedom—though part of the natural order and a natural right—has not always been recognized and thus has not always been operative in an economy. As we shall see, the development of commerce advances the natural "fact" of liberty, the fact that each of us can best look after our own affairs, by making possible the independence and the division of labor, free trade, and complex commercial exchanges. However, Smith attaches an important proviso to his notion of natural liberty. For him, natural liberty pertains to personal and economic liberty, the liberty to conduct one's personal affairs. This liberty to "better one's condition" should not be confused with democratic freedom, however: Smith was both an aristocrat and a pessimist in political affairs. He had little faith in commoners' abilities to govern themselves or to make sound political judgments that were not merely in their self-interests and the interests of their friends and cohorts. He found such democratic governance to be both ineffective and detrimental to peace and public safety.[25] Nor does he argue that political freedom depends on, is derived from, or is a function of economic freedom. More will be said about the question of democratic freedom in Chapter 6.

## Property Rights

Rights not to be physically harmed, rights to personal liberty, and rights to the preservation of one's reputation are natural rights. They do not depend on conventions, the economy, or society. However, following Hume, Smith contends that "estate" or property is an acquired right. Property rights, unlike natural rights, are not rights that one possesses simply because one is a human being. Because one possesses or acquires property does not thereby imply that one has a natural right to it, because property, unlike liberty or reputation, is socially defined and can be transferred to and from others.[26] The definition and scope of property and what counts as a valid property acquisition or exchange evolve through history, and because the ownership of property creates inequalities, property requires the institution of government to be protected.

Under "estate" or property rights, Smith distinguishes real rights and personal rights, the former being those rights to one's immediate possessions and the latter are what is due as a result of a contract, loan, and the like. Real rights are exclusive property rights; that is, they are entitlements against other persons or possible claimants to that property by reason of the validity of the property claim. Real rights include not only property per se but also servitudes, those claims on property that one must cede or must be ceded to one, such as right of ways, pledges or mortgages, and inheritances. Personal rights are those rights to properties due to one personally as a result of contracts or other personal arrangements.[27]

In the *WN* in reference to regulations concerning ownership of mines that one discovers and works, Smith writes: "In both regulations [of tin and silver mines of which the discoverer becomes the proprietor even if it is on another's land] the sacred rights of private property are sacrificed to the supposed interests of publick revenue."[28] One of the questions that arises when reading this passage is what Smith means when he refers to "the sacred rights of private property." Given what he says in the *Lectures* about the adventitious nature of property rights, it makes sense to interpret this passage as saying that rights to validly acquired property are sacred in the sense that having defined these rights under law, it ordinarily is unjust to violate those rights even in the public interest. But Smith cannot mean that property rights are sacred in the sense of being natural rights, because he argues throughout the *LJ* that this is not the case.

Sometimes Smith appears to adapt John Locke's thesis that one has a natural right to the property that is a result of or is improved by one's labors, that is, the "fruits of one's labors." If this is the case, then Smith would appear to ground property rights in a natural rights–based labor theory of property. For example, he says: "The property which every man has in his labour, as it is the original foundation of all other property, so it is the most sacred and inviolable."[29] In this context he is talking about problems with the institution of long apprenticeships as a form of semienslavement that prohibits laborers from getting a decent wage or from changing employment. Smith means to justify the rights of laborers to independence from the indenture of oppressive apprenticeships and to support his arguments elsewhere in the *WN* that laborers have rights to a decent living wage.[30] It also illustrates how he uses both descriptive and normative techniques to make his points. In the *LJ* Smith describes the origins of property and distinguishes property rights from natural rights. In this particular passage in the *WN* he defends the rights of apprentices and makes a normative claim on their behalf.

Because Smith writes that "labor . . . is the original foundation of all other property," however, Robert Boynton Lamb argues that Smith, allegedly following Locke, has two theories of property. Lamb points out that Locke defends property as a natural right in the state of nature and superimposes on that a theory of civil property. In brief, from the natural rights to life, to one's body, and to freedom, one has a right to labor, according to Locke, and because the labor is one's own, one has a right to the "fruits of one's labor." Locke also argues that in a civil society formed by social contract out of a state of nature, the introduction of money and conventions governing property rights allow the expansion of private property beyond these "fruits."[31] Scholars have long recognized the difficulty of justifying civil property from Lockes' theory of natural rights and, in particular, from a labor theory of property. This leads Lamb to conclude that Locke has two theories of property, a natural rights theory and a theory of civil property. Lamb finds that Smith, too, following the precedent of Locke, holds a similar view, that is, a natural rights–labor theory of property and a civil theory of property, two theories that Lamb finds to be irreconcilable.[32]

Lamb bases this claim on Smith's statement in the *WN* quoted in the previous paragraph that the property of labor is a sacred right, and because of Smith's oblique reference to the "sacred rights of private property." Lamb also appeals to what Smith says about the origins of civil property in the *LJ* to justify his interpretation. In tracing the origins of property and civil government, Smith outlines four stages of development from primitive society to the age of commerce. In the first stage, the stage of hunters, there are no explicitly defined property rights, although there is an understanding that hunters have the rights to keep their "kill." In the second stage of development, the stage of shepherds, the shepherds develop a notion of private property through the ownership of their herds. In both stages the hunters and the shepherds have rights to the fruits of their labors. In the second stage this right is expanded when the herds multiply, thus leading to the distinction between people with property and those without property. In the second stage, and more fully in the third and fourth stages, property is developed beyond the "fruits of one's labors," and because not everyone has property, it becomes necessary to define and enforce rules for the acquisition, transfer, and security of one's possessions by the civil government. In these two early stages Lamb sees Smith, like Locke, as holding a natural rights–based theory of property, a theory that becomes problematic with the advent of civil property, in which the possibility of property extends beyond one's labors.[33]

Lamb presents an interesting reading of Smith. It is surely true that Smith was influenced by Locke and that, at least in the *WN* passage just cited, he talks as if labor were the origin or "foundation" of property. Yet while arguing that labor is the property of the laborer, Smith does not develop a labor theory of property. Even when he discusses the productivity of the "fruits" of labor, he is consistent in his arguments that property is an adventitious right. As I shall argue in Chapter 5, although labor belongs inviolably to the laborer, the right to the productivity of one's labor as property or a commodity is an acquired right, according to Smith.

Interestingly, in a book on Locke's theory of property, John Tully questions an interpretation such as Lamb's.[34] Tully's analysis is helpful in deciphering what Smith meant by property and property rights. Tully points out that Locke uses the term *property* in a number of senses. First, in general, properties define the powers of ownership, that what is at issue cannot be taken away without consent. Some properties such as person, life, liberty, and self-preservation are inalienable properties that cannot be taken away even with full consent. These are what Smith refers to as natural rights, but he does not use the term *property* in this context, for good reason, I shall suggest. Other properties are alienable and can be transferred, with consent. Concerning the latter, Locke appears to distinguish two sorts of property ownership. Inclusive or communal ownership of common property gives one the right not to be excluded from this common property and, indeed, a limited right to use such property for self-preservation. Exclusive ownership defines private ownership to material properties which includes the right to exclude others from one's property.[35] Smith, following Hume's example, adopts the latter definition for property and does not dwell on rights to use common property. In fact, Smith believes, it is only when people make claims to exclusive ownership that the notion of property and property rights comes into existence.

Tully explains further that Locke does not have a labor theory of property: "Labor justifies neither the accumulation of nor rights over one's goods."[36] Rather, Locke's point is that one has natural and inalienable rights to one's life, body, liberty, and survival, that is, the fulfillment of basic needs. One, but only one, of the means to survive is to labor, and one has rights to the fruits of one's labor not because of the laboring but because this is necessary for preservation. This also explains Locke's famous proviso for property accumulation in the state of nature, that one should take only what one can use, with enough left over for others.[37] Tully points out that according to Locke, because laboring is not the only means to survive, charity is also a "natural duty" for those with ability and property left over after their own needs are satisfied. So, Tully concludes, according to Locke one does not have a natural right to property as a result of one's labor. Rather, a "property in something is the completion of man's natural right to the means necessary to preserve and comfort himself and others. . . . [F]ixed property in land does not have a natural foundation."[38]

Tully's conclusion is that exclusive rights to material and alienable properties are conventional rights, according to Locke. As properties are accumulated beyond what one can use, with enough for others, civil society is formed not only to safeguard natural rights but also to protect common property and inclusive rights against encroachment by owners of exclusive properties. Locke sees government as the guardian of common property and the watchdog to limit exclusive property claims, rather than the protector of unlimited acquisitions. Rather than the precursor of capitalism, then, according to Tully, Locke is firmly in the mercantile tradition.[39]

Because Smith claims throughout the *LJ* that property is an acquired right, it is consistent to conclude that he, like Hume, may have interpreted Locke as having two theories of property and saw the problems therein, problems that Smith hoped to avoid by focusing on exclusive ownership and by declaring that property rights are acquired rights, the notion of which develops only in civil society. Or like Tully much later, Smith may have read Locke not as having a labor theory of property and thereby adopted Locke's thesis that exclusive property rights are conventional rights.

In any event, Smith did not adopt Locke's multiuse terminology of the word *property*. Instead, he uses the term *property* to refer to alienable exclusive ownership claims. This terminology is much less confusing than Locke's and avoids the problem of talking about the inalienable property rights to one's life, body, and liberty. Smith, following Hume, does not distinguish between common or inclusive property and exclusive property. Hume argues that in a state of nature with no scarcities and given the idyllic condition (never realized) of perfectly benevolent human beings, the question of property claims would not arise. In a state of abundance, common property and unrestricted access preclude the necessity of such distinctions. In such a context, too, justice would be unnecessary. It is only because of the dominance of self-love, the "avidity" of human nature, and natural and human-made scarcities that human beings—who cannot live outside society—adopt conventions that regulate the acquisition, possession, and transfer of various properties. Justice, then, is "artificial" in the sense that it is both conventional and utilitarian, although it "naturally" arises in any social context because of the self-interested condition of human nature.[40]

According to Smith, because property is an adventitious right whose scope is defined by society, the notions of property and property rights come into ordinary parlance only when a society advances to the stage of delineating possessions, acquisitions, and trades. At that stage there develops an inequality of possessions, and both the inequality and the subsequent scarcity create a need for civil government to protect and stabilize property claims: "The acquisition of valuable and extensive property . . . necessarily requires the establishment of civil government."[41] Therefore, according to some interpreters of the *LJ*, following Hume's arguments, Smith implies that economic development is a necessary condition for civil government, law, and laws of justice. Smith's argument, however, is slightly different.

According to the *TMS*, as part of the natural order, natural jurisprudence is the basis for any society. So underlying property, avidity, scarcity, and even economic progress is natural jurisprudence which grounds particular laws governing economic relationships and thus property and property rights. As grounded in natural jurisprudence, justice is not merely a conventional development. Sometimes, too, Smith talks as if it were when a society developed rules for adjudicating claims that the discrimination of possession, acquisition, and trades of property evolve, because there is a means in place for justifying property acquisition and accumulation and for settling conflicts. Moreover, unlike property, justice is not simply utilitarian. Rather, natural jurisprudence pertains to the virtue of impartial evaluation and adjudication, and laws of justice are to reflect that perspective. Property rights, property laws, and civil government arise from what Smith calls the "natural progress which men make in society,"[42] much of which is economic and social progress nevertheless based on an underlying natural order. So the relationship between economic progress, the development of the notion of private property, and the evolution of civil government is complex, and it is not clear that one factor necessarily precedes or is a condition for the others.

Although property rights are acquired rights, property rights are nevertheless perfect rights, according to Smith. That is, one has a right to validly acquired property, and so it is the perfect duty of others not to violate that right, and the duty of the government to protect it. Thus property rights are "sacred" in the sense of requiring perfect duties.[43] But Smith recognizes that in different societies and at different stages in history, what counts as a valid acquisition, and thus a perfect duty to prevent a violation of that right, varies. He claims that what counts as a violation of a property right in any society is determined by what an impartial spectator would judge to be such a violation. Because the impartial spectator is an average person disengaged from a situation in order to evaluate it dispassionately, the definition of duties to protect property in terms of the judgments of an impartial spectator accounts for both an impartial evaluation of a possible violation of a property right and the fact that in different historical periods and in different societies what an impartial spectator deems to be appropriate property acquisitions, transfers, and their violations may well vary considerably.[44]

Smith lists five sources of property acquisition by which I assume he means valid sources of acquisition, that the rights to property so acquired must be protected. These include (1) possession or occupation, the acquisition of property simply by possessing it (e.g., picking an apple, staking out a land claim) or by work. Property is also acquired (2) through accession—for example, the rights to the minerals of land one owns, the milk of one's cow or even the fruits of one's labors—(3) by prescription, in which property is deeded to someone because he or she has possessed it for some period of time, (4) from succession or inheritance, and (5) by voluntary transfer in which the individual making the transfer intends to make the exchange and the property is actually delivered to another party.[45] In each of these sources the acquisition or exchange is valid if an impartial spectator does not deem the acquisition or exchange a violation of rights. What Smith lists as three other forms of real property—servitudes, pledges, and privileges, as well as personal property rights—also can be considered valid if they pass the test of an impartial spectator.[46]

The introduction of the impartial spectator distinguishes property rights from other perfect rights. In the case of natural rights one has perfect duties to protect the nonviolation of these rights in every circumstance and in all societal and historical conditions. In the case of property rights, the extent and scope of these rights are conditioned by the societal and historical situation. The impartial spectator in that context, through his or her judgments, formulates the general rules, in this case, the general rules governing property, its acquisition, succession, and transfer. It appears that the extent and scope of property are culturally relative, but by stating five valid sources of property acquisition and transfer, Smith presupposes an underlying uniformity of human nature in which we all would accept these five sources as exhausting the format of proper acquisitions and exchanges, although the extent of property rights and rules governing property are subject to cultural and historical idiosyncracies.

Smith sees no problem with his claim that property is a perfect but adventitious right whose violation is determined by an impartial spectator. He appears to contend that because property rights are individual rights that traditionally have had the status of natural rights, they are perfect rights, but he does not defend this contention. Rather than adopting Locke's two theories of property, then, Smith appears to have conflated the two. Although property is not a natural right, a violation of that right—like the violation of the right to the fruits of one's labors in Locke's state of nature—entails perfect duties to correct that violation. Yet unlike other perfect rights, the determination of the violation is left to the impartial spectator. But how can one have perfect duties to redress violations of historically and culturally determined rights if the kind of duty allegedly entailed is defined by the scope and extension of that particular right? Thus while trying to resolve what is commonly called Locke's "two theories" problem, by declaring that property is an acquired right, Smith has not succeeded in adequately explaining how a perfect duty to protect a property right is historically relative and yet perfect.

## The Origins of Property and Government

Smith envisions the development of property, labor, land, and civil society as commencing with the distinction between those who have property and those who do not, a distinction that gives rise to the necessity of government. In turn, the development of markets, the valuing of labor as a commodity, and the notion of economic liberty evolve from the institution of property. Subsequently, these phenomena lead to the division and specialization of labor and thus economic growth. In the *LJ* and *WN*, Smith traces the "natural progress" of property rights along with the necessity for civil government and laws of justice through four stages. These stages indicate economic development as well, and it is evident that Smith thought that the phenomena of private property, property rights, and civil government are closely interrelated to the economy of a society. Although he sees the foundation of human development—the natural order or human nature—as static throughout human history, a thesis he carefully develops in the *TMS*, it is clear that in the *LJ* and *WN* he finds these stages of economic and civil development to depict what later writers called evolutionary progress. The obvious differences in description between human development and economic and civil evolution, according to Jacob Viner, are evidence of irreconcilable differences between the earlier *TMS* and the *WN*. What is interesting, however, is the fact that throughout his writings, Smith finds a natural order, expressed in terms such as "natural liberty" and "natural jurisprudence," underlying all forms of human society. In addition, as Macfie points out, general (moral) rules are derived from (changing) experience and thus participate in this evolutionary process as well.[47] So perhaps it is more consistent to argue that although not abandoning the notion of a static or stable natural order, Smith recognizes that human beings develop and change, interpreting and embodying to a greater or lesser extent elements of natural rights and natural jurisprudence at different stages of history and economic sophistication. Finally, although it is true that he finds the fourth state, the age of commerce, the most advanced economically, he does not see the four stages as merely linear progression, nor does he imagine that history cannot reverse itself at any stage.[48]

According to Smith, the first stage of the development of a civil society is the age of hunters. Because hunters have little property except what they kill for their own consumption, the economy is a simple one. Hunters simply keep or trade the "fruit" of their hunting. Because no one owns any substantial property and because no one is economically dependent on others, there is no need to define property rights or to establish a complex system of justice. The government is simple and participatory, and people are able to express their natural liberty individually and democratically, as there is little utility or need for excessive authority or demand for obedience. "Thus among hunters there is no regular government; they live according to the laws of nature."[49]

The second stage, the age of pasture and of shepherds, is more complex because people begin staking out claims to property in the form of animal herds and grazing lands. Such claims create property as a result of occupation, transfer or trade, and accession. Notice that although Smith does not adopt a natural rights–based labor theory of property, he sees property rights as arising in the first instance from what one does, or one's occupation, if that also includes or results in the possession of property. What is interesting, however, is that although the notion of property as land or possessions comes from occupation, the concept of labor as a commodity arises only in the fourth state, the stage of commerce. With property there are inequalities between those who own property and those who do not, and therefore there is a need for some sort of order to protect property from nonowners. About this stage Smith is quoted as saying: "The appropriation of herds and flocks, which introduced an inequality of fortune, was that which first gave rise to regular government. Till there be property there can be no government, the very end of which is to secure wealth, and to defend the rich from the poor."[50] The rudiments of a judicial system are formed to settle property claims and to protect owners from nonowners, and the beginning of civil government is founded on the authority of a chieftain to whom obedience is both required and necessary to protect one's property. This, too, is the beginning of a monarchical form of government. Moreover, it is in this stage, with the advent of property, that lines of property inheritance are established.[51] Government is necessary, and laws of justice need to be established, then, according to Smith, when a society has progressed to the point of acquiring property, not because of any natural or original consent or contract, but rather, in the first instance, to establish authority for the protection of property rights.

The third stage, the stage of agriculture brought on when people begin to live in settlements, has at least three substages. First and most obviously is the stage of farming, in which property is further defined in terms of ownership and use of land. Even in these early stages Smith found rudimentary forms of what he took to be the three distinct powers of government, the legislative, the judicial, and the executive.[52]

The second substage of the stage of agriculture is the formation of city-states. These cities were quasi-independent, paying a baron or lord for protection, as there seldom was a central nation or the national authority of a sovereign or monarch. This is also called the *allodial period*, in which land is privately owned and large landholdings are created. Tenants or serfs who lived and worked on the land were dependent on the baron or lord for protection and sustenance, to the point that most of these tenants lived in a state of semislavery. When land was traded or sold, the serfs were part of that negotiation. Barons therefore often had their own armies to protect themselves and their lands and serfs.

The third substage is the age of feudalism. With the establishment of nations and sovereigns with central authority, a new relationship was created. Landholding lords looked to their monarch for protection, and conversely, the monarch's fear of losing the lords' obedience led to the establishment of feudal arrangements. Lords became the hereditary owners of land under the monarch, and any sense of city-state democracy was lost. Tenants became even more dependent on their lords for security, but at the same time they were given life leases on property for which they paid a high rent. At the same time, according to Smith, during this period of absolutism, peace and order allowed the development of parliaments and a legal system that formed a secure foundation for the administration of justice.

The overindulgence of the feudal lords, their increased demand for consumer goods, and the independence of the city-states, Smith believed, led to the collapse of feudalism. The demand for luxuries led the lords to pressure their tenants to be more productive. They dismissed the weak tenants but competed for the productive tenants. The productivity of the tenant, then, rather than the tenant qua tenant became valued as a commodity. The value of labor's production as a commodity, as distinct from the laborer, was, for the first time, recognized, and so desirable (productive) tenants were able to bargain for their labor. Thus agricultural labor became both more efficient and more independent. With the existence of leases to protect the tenant against the arbitrary authority of the landlord and the valuing of his labor, the tenant could sell the productivity of his labor as a commodity without selling himself, and thus he was freed from his identification with it. "Even a tenant at will, who pays the full value of the land, is not altogether dependent upon the landlord,"[53] Smith argues, because the landlord could not demand more than the lease required. Thus the tenant—as he became independent from the landlord and distinct from his productive labor that he could barter—could nurture the idea of personal liberty.

Accompanying the breakdown of feudalism was the rise of independent towns that were granted self-governing privileges by the monarch in return for their allegiance against feudal lords or the church. With this independence, the towns began their own commerce and trade, and the demand for goods both by the feudal lords and townspeople led to foreign trade. This in turn encouraged domestic manufacturers to enter the market to satisfy this demand, thus providing new jobs in what used to be primarily an agrarian economy. The development of foreign trade and employment, along with the advent of private commerce among tenant farmers and between farmers and townspeople, led to the fourth stage, the age of commerce.[54] Whether trade, manufacture, the valuing of labor as a commodity, and the new development of cities led to the breakup of feudalism or, alternatively, whether the breakdown of feudalism led to commerce and manufacture is uncertain. Smith argues for the former in the *WN* and talks about both in the *LJ*. Similarly, whether the market for productive labor transformed productivity into an independent exchangeable commodity or, alternatively, whether the independence of labor and the valuing of productivity as a commodity transformed the economy into a commercial society is also unclear. What is important to Smith is the development of labor and commerce in which various sorts of productivity and properties are exchanged, allowing for the possibility of economic independence within a constitutional monarchy. At this stage the results of the evolution of private property produce well-established property rights and a judicial system to govern those rights, a system developed in the third stage, together with the valuing of labor and the possibility of free commerce. Eventually this permits the division and specialization of labor which are crucial to economic growth. The rise of commerce, however, does not indicate a return to the democracy experienced by hunters, nor does Smith find this to be desirable.[55]

We see, then, the development of civil society with a strong government whose authority is established through a lineage of respect and whose obedience is required to protect property rights. Government develops as a means to protect property and property rights for, as Smith says in the *WN*, "wherever there is great prosperity, there is great inequality. . . . The acquisition of valuable and extensive property, therefore, necessarily requires the establishment of civil government."[56] In the fourth stage of history, the rise of commercialism is accompanied by a sense of economic independence and the realization of natural liberty in economic affairs. According to Smith, private property, economic liberty, the valuing of labor as a commodity, and market exchanges all are highly desirable conditions for, and the result of, commerce. There are at least three reasons that this is the case. First, economic liberty most closely emulates the natural order.[57] Second, as Locke and Hume implied, the development of private property is essential to the division of labor and thus economic growth. For if a laborer keeps the full fruits of his labors, there will be no surplus capital for large investments, and an economy will scarcely be able to progress beyond the hunter or the primitive shepherd stage, as there will be little capital accumulated for investment. Third, these phenomena provide the conditions for the development of commerce and ultimately economic growth because—as I shall examine in more detail in Chapter 5—only with the division and specialization of labor, a division dependent on the valuing of labor as a commodity, the accumulation of capital, and unsatisfied market demands for goods can the age of commerce lead to societal economic well-being.

Smith is often cited as the intellectual predecessor to Marx, setting out the framework for Marx's analysis of history. Smith is usually credited with the idea that the economy of a society forms the basis for its political, social, and even philosophical superstructure. Along with Locke he may have initiated the labor theory of value that Marx later adopted and even provided the groundwork for Marx's notion of alienation. Because Smith's four stages of economic and civil progress represent stages of economic development whose character is the foundation for the notion of private property, property rights, and thus the economy, Smith is sometimes read as saying that society and civil government are defined by, depend on, and exist only as a result of the underlying economic base. Each stage is defined by a particular mode of subsistence that establishes the material conditions for that society's character and culture. Smith is then regarded as a historical materialist who, seeing regularities in history, depicts society as a gigantic social machine, for which the economy is the driving force. At each stage the economy of a particular society determines its political, civil, and social structure and direction and sets the conditions for the next stage. According to this materialist interpretation of history, any economy must progress through these four stages. The development of property and property rights are necessary for government and the valuing of labor, all of which form the conditions for the beginning of commerce. The age of commerce is the ideal, for it is only at this stage that an economy achieves economic freedom, industrialization, and thus economic growth. Accordingly, the stage of commerce is the final stage in economic progress, because it is then that Smith's ideal of a truly free commercial economy can be realized.[58]

This interpretation of Smith, however, may be an overreading of what he meant to accomplish with his four stages. Although it is true that Marx read and was greatly influenced by Smith, one must be careful not to read back Marx anachronistically into Smith. Smith's four stages are meant to represent a process of civil and economic development rather than historical accuracy, and Smith would not find it strange to discover a society that did not progress along these particular lines. He would see himself as merely describing how, in fact, civil society has usually progressed. He found the development of a commercial society with property, economic liberty, and a framework of justice crucial to economic growth and economic liberty, because the division of labor, dependent on property, market exchanges, and the valuing of labor, provides a necessary condition for economic growth. But Smith could imagine that the age of commerce could also arise through other kinds of developmental stages. For example, economic growth could take place in a regulated economy such as Smith found in eighteenth-century England, although he would argue that such an economy is unjust and less efficient than the one he envisioned.

As I stated earlier, in counterposition to Hume, it is not clear that Smith finds the advent of private property to be the determining factor in the progress of civil government. He does say that "till there be property there can be no government,"[59] and government is essential to preserve property rights. He also writes: "Property and civil government very much depend on one another. The preservation of property and the inequality of possession first formed it [government], and the state of property must always vary with the form of government."[60] The hunters have a modicum of government and a rudimentary idea of justice, even though they have a simple economy with only poorly defined property rights, and that modicum of government and justice serves as the basis for civil development, necessitated by the advent of property. Although the expansion of civil government is triggered by property arrangements, the kinds of arrangements that flourish in turn depend on the civil framework. As we shall see in Chapter 6, whereas the stage of commerce opens up the idea of economic liberty, commercialism, and trade, the doctrine of mercantilism that pervaded eighteenth-century English economic thought justified regulations on the economy of that day that prohibited certain property arrangements that Smith would find propitious for the ideal political economy. Although private property gives rise to a complex civil government, Smith claims, the laws of justice of any civil government should be based on principles of natural jurisprudence. Civil government is not merely derived from property but is also grounded in natural jurisprudence. Therefore Smith does not imply merely that property gives rise to government; the relationships between an economy and its institutional framework of government are much more complex.[61]

Finally, it is not certain that the age of commerce is the "final stage" for Smith. He finds faults even in a system of perfect liberty, faults that are endemic to that system,[62] and the age of commerce need not necessarily herald economic well-being. Thus Smith is not an economic determinist, despite his influence on Marx and Marx's historicism. In Chapter 5 I shall describe the influence of Smith's theory of labor on Marx's idea of alienation.

## Civil Rights

Property rights are embedded in civil society, and in addition to personal rights Smith distinguishes two other kinds of acquired rights, the rights that one has as a result of the kinds of injuries one may receive as a member of a family, and the rights one has as a citizen of a state. Concerning family rights, one point is important to understanding the *WN*. When talking about family rights Smith separates the rights of a husband and wife, of a parent and child, of guardianship, and of a master and servant. During Smith's time—and indeed until 1980 even in this country—the term *master–servant* was used in the law not only to describe family or servant relationships but also as a general term to refer to any sort of employment relationship.[63] In the *LJ*, however, Smith concentrates primarily on family, servant, and slave relationships. Thus what he says there is not helpful in interpreting his claim in the *WN* that "the property which every man has in his own labour, as it is the original foundation of all other property, so it is the most sacred and inviolable."[64]

Smith's third division of rights, the rights of a citizen, distinguishes between the rights of a sovereign against the citizens and the rights of the citizens against the sovereign. Smith's justification for obedience to the government is not based on the idea of an original or social contract. Like Hume, in a number of places in the *LJ* Smith is critical of that idea, particularly as Locke formulated the notion. Interestingly, Smith does not attack the social contract theory on the grounds that it assumes an undue degree of self-interestedness as a motivating condition for forming a contract. Nor does he consider that the postulation of a social contract might be a heuristic device for getting at the basis and nature of human agreements.[65]

Rather, Smith takes the social contract idea fairly literally. He does not see the normative implications of a social contract, nor does he appear to imagine that the idea is not an actual description of an existent state of nature. That human beings, by consent, create a government out of a state of nature seems unlikely. The notion of limited obedience and the process of making a contract are historically inaccurate, Smith argues.

> This tacit contract, so much talked of, is never thought of either by the governors nor the governed. The principles on which this allegiance [to the governors] are really founded are those of authority [and] of public or generall utility.[66] . . . All have a notion of the duty of allegiance to the sovereign, and yet no one has any conception of a previous contract either tacit or express.[67]

The process of contract or consent questions the authority of the sovereign or government and is too burdensome for governance. Smith believes that government is created not from consent but from "natural progress" or social evolution. This precludes even the hypothetical possibility of a social contract. Government is necessary to protect property and was originally based on authority so that such rights were adequately protected without question. Our obedience is based on that tradition of authority. Smith traces authority to wealth, not to the dependence of the poor on the rich but, rather, to the fact that we tend to admire and respect people of wealth and distinction.[68] Obedience is also based on the recognition of the utility of living in a society and being protected by an orderly government.[69]

Smith also views the rights of citizens as vague, especially the right to citizenship itself. Adopting what was even in Smith's time a commonplace division of government, he claims that the rights of citizens are best protected when the power of government is divided into three distinct branches: the legislative, the executive, and the judicial. But the authority rests in government, not with the citizens. The sovereign, being the natural and traditional authority, is absolute, and treason is justified only when the sovereign exhibits "absurdity and impropriety of conduct and great perverseness."[70] Although Smith asserts that "whatever be the principle of alledgiance, a right of resistance must undoubtedly be lawfull, because no authority is altogether unlimited,"[71] he also remarks that "no government is quite perfect, but it is better to submit to some inconveniences than make attempts against it."[72] Smith's conservatism is based on a fear of anarchy and his belief that peace, law, and order in the long run best serve economic and civil progress. This recognition of the value of peace coupled with his skepticism about the governing abilities of the common people, not his love of monarchy, account for his pessimism about a democratic process, although such pessimism does not deter his enthusiasm about economic freedom.

Though recognizing his arguments against the idea of an "original contract," a few commentators contend that Smith believes that there is an implicit social contract between government and its citizens based on their lawful right to resistance when the government gets out of hand.[73] Smith, however, sees the right to resistance as based on the natural right not to be harmed and thus, derivatively, the right to resist personal injury and also as based on his recognition that authority and obedience to a human government cannot be unlimited. He does not regard the right to resist as evolving from a social contract between citizens and government. Rather, because justice proscribes harming another, one has a right not to be harmed and thus a right to resist when that harm is overwhelming. But Smith finds that seldom is harm grave enough to warrant resistance, because revolutions usually create more harms for citizens. One might read into this an implicit social contract, but it is clear that this was not Smith's intention.[74]

## Natural Jurisprudence

Smith begins the *Lectures on Jurisprudence* by defining jurisprudence as follows:

> *LJ(A)*: Jurisprudence is the theory of the rules by which civil governments ought be directed.[75]
> *LJ(B)*: Jurisprudence is the theory of the general principles of law and government.[76]

These "rules" or "principles" include laws of justice whose object is "the security from injury"[77] and "maintaining . . . perfect rights."[78] Smith's notion of jurisprudence in the *LJ* derives from his theory of natural jurisprudence developed in the *TMS*, and in both texts he claims that justice originates in the natural order. To review, in the *TMS* Smith argues that the moral sentiment of justice derives from the social passions. Justice, as distinct from benevolence, arises from the impartial social interests and, in that respect, is the "consciousness of ill-desert."[79] As the virtue of fair play, justice is the basic virtue. The just person "would act so as that the impartial spectator may enter into the principles of his conduct."[80] So it is evident that justice is a necessary condition for self-command. Fair play is not unimportant to economic affairs, and in the *WN* Smith writes that "it can never be the interest of the proprietors and cultivators to restrain or to discourage in any respect the industry of merchants, artificers and manufacturers."[81] Thus the norm of fair play, a norm derived from the social passions and therefore "natural," Smith explains, should govern both one's conduct in economic exchanges and any regulation of that conduct. Linking this idea to his statement in the *WN* that natural liberty is restrained by "laws of justice," that is, by the precepts of natural justice to act fairly and not to harm others, and to his admonition that the protection of some interests at the expense of others precludes "equality of treatment," the rational economic actor begins to look like the impartial spectator whose liberty is both self-restrained and restrained by laws of justice not to harm others, not to encourage practices that result in inequality of treatment, and to practice fair play. Indeed, Knud Haakonssen suggests that "what gives unity to Smith's theory [of jurisprudence] is the spectator approach." This interpretation may be reading too much into the texts, however, because in the *WN*, Smith speaks little of the impartial spectator or the virtue of justice. Rather, he focuses on laws of justice, principles to maintain rights and to protect citizens from harms.

Laws of justice are based on natural jurisprudence, for "every system of positive law may be regarded as a more or less imperfect attempt towards a system of natural jurisprudence."[82] As the basis for law, natural jurisprudence includes three principles: (1) the negative harm principle, that each of us has a right not to be harmed physically, by reputation, or concerning our personal liberty; (2) the fair play principle, that each person has an equal right not to be "justled" or given an unequal opportunity or unfair advantage; and (3) the perfect rights principle, a principle derived from the *LJ* to protect both property rights and natural rights, that every person has duties not to infringe on the perfect rights of others. As society progresses through these four stages, the laws of justice are formulated and fine-tuned to protect property rights and settle contract disputes as well as to protect natural rights, and throughout the *LJ*, Smith demonstrates how natural jurisprudence can be spelled out in law.

Nowhere, in any of his writings, does Smith draw up a notion of distributive justice. This omission is deliberate, he argues in the *LJ*, because justice pertains only to perfect rights and their corresponding duties. Only perfect duties apply equally to everyone and therefore can be codified and enforced. "The very definition of a perfect right opposed to the offices of humanity, etc., which are by some called imperfect rights, is one which we may compell others to perform to us by violenc[e]."[83] Because one has perfect duties not to harm another, not to treat another unfairly, and not to infringe on his or her rights, laws of justice to enforce these duties are universally applicable to every person, impartially and equally. On the other hand, Smith maintains, any form of distributive justice is derived from imperfect rights and therefore is neither universally applicable nor mandatory.[84] In addition, he implies that most forms of prescribed distribution are unfair because "to hurt in any degree the interest of any one order of citizens, for no other purpose but to promote that of some other, is evidently contrary to that justice and equality of treatment which the sovereign owes to all the different orders of his subjects."[85] Recalling that in the

*TMS* Smith declares that justice is the only enforceable virtue, it is commutative justice that fits this qualification.

In the *TMS* justice is both one of the social virtues and, according to Smith, the basis for a viable society without which social relationships would be impossible. Thus justice has a crucial utilitarian function, and this becomes more evident in the *WN*, in which justice plays a central role. Smith is repeatedly critical of laws, regulations, or individual economic actions that are unfair to some participants or group of participants in an economic exchange. Smith states over and over that most restraints on natural liberty are bad. On both the individual and the political level, the "system" operates best when it contains a minimum of regulations, laws, and restraints. For example, Smith argues that people should be free to live where they wish, to trade or do business as they please, and to accept or change occupations at will.[86] Yet, even in these passages it is clear that he does not advocate a form of anarchy, because he links natural liberty with natural jurisprudence.

In Book V of the *WN* Smith observes that "commerce and manufactures can seldom flourish long in any state which does not enjoy a regular administration of justice."[87] Justice has to be codified into law and enforced, according to Smith, for even though most of us do not deliberately set out to harm one another, few of us internalize the ideal of fair play. Although we cooperate with one another in economic affairs, this cooperation often involves group collusion against others. For example, traders often fix prices; manufacturers or merchants jointly devise plans to exploit customers; or a consortium of employers may exploit laborers.[88] Laws of justice thus are essential not only to restrain self-interests but also because we are naturally cooperative. Smith is not revising his view of human nature developed in the *TMS*, that each of us has social as well as selfish passions and is capable of sympathetic understanding of another. Instead, as we shall see in the next chapter, he finds both the social and the selfish passions operative in economic affairs as well. Although some of us bridle our self-interests and social attachments with an internal sense of fair play, this is not universally the case. Thus Smith gives to the sovereign the "duty of protecting, as far as possible, every member of the society from the injustice or oppression of every other member of [it],"[89] even to the extent of restricting natural liberty when the "security of the

passions, laws of justice are necessary to prevent harm and ensure fair play.

Despite the important utilitarian roles of justice in economic and political affairs, justice is not merely an "artificial" or useful virtue, as Hume claimed,[91] nor are laws of justice merely utilitarian. Smith agrees with Hume that laws of justice are necessary in social relationships and in that sense arise because human beings are naturally and by necessity social. Smith, then, acknowledges the utility of justice in both political and economic affairs. But he also notices that we appreciate the beauty of justice and abhor injustices even when the utility or perceived relevance is unclear. We also protest injustices even when we are not acquainted with those involved or when the individuals at issue are reprehensible,[92] because justice arises from the social passions and is rooted in the natural order whose normative standards are not merely utilitarian. As Richard Teichgraeber perceptively points out, however, Smith's notion of justice is different from that of the natural law theorists.[93] The virtue of justice does not instill people with civic or political virtues, so that one's first commitment is to society. In fact, virtues of "system" should be discouraged. Rather, justice is a negative precept, and laws of justice are ordering principles that restrain harms to private interests while allowing as much individual liberty as possible. Smith admires the citizen of the world whose "interest of this great community he ought at all times be willing that his own little interest should be sacrificed."[94] Natural benevolence and a public spirit are admirable individual virtues, but only the respect and promotion of negative justice are required.[95] Smith, then, does not belong in the natural law tradition, and he is not merely a conventionalist regarding justice, because natural jurisprudence, the basis for justice, exhibits itself imperfectly, although in its proper form, negatively, in conventions and laws of particular societies.[96]

How do laws of justice relate to natural liberty? Smith writes that "every man, as long as he does not violate the laws of justice, is left perfectly free to pursue his own interest his own way, and to bring both his industry and capital into competition with those of any other man, or order of men."[97] At least one commentator who holds a form of the "self-interest" reading of Smith discussed in the Introduction interprets this passage to mean that ideal "system" of natural liberty is one in which each "atomic, rational, self-

any other man, or order of men."[97] At least one commentator who holds a form of the "self-interest" reading of Smith discussed in the Introduction interprets this passage to mean that ideal "system" of natural liberty is one in which each "atomic, rational, self-interested individual,"[98] in isolation with others, has the liberty to do as he or she pleases. But to contend that Smith holds a radical individualistic view of economic activity in the *WN*, in which "atomic" free self-interested individuals acting in "unrestrained self-interest" is the norm for economic behavior, not only ignores arguments in the *TMS* but also belies what Smith says in the *LJ* and *WN* about justice and its central relationship to liberty. Laws of justice, to extrapolate from the *TMS*, reflect the natural jurisprudence of fair play. In the natural order individuals are both free to pursue their interests and, as social beings, self-restrained by the desire for the approval of others, which is why Smith refers to this order as a "system of natural liberty." Such a system, restrained by justice, is closer to the natural order than is anarchy or the license to do whatever one pleases. A society that emulates the system of natural liberty is most harmonious and closest to the natural order.

To summarize, because justice comes from the natural order and is the framework for a political economy, one can think of justice as functioning on three levels. First, natural jurisprudence, the "respect for what is due" comes from the social passions; it forms the natural and normative ground of our idea of justice; and it is the basis for any society. Second, the principles of commutative justice—not harming another, respecting perfect rights, and encouraging fair play—are perfect duties derived from the natural order and adjudicate violations of perfect rights. Third, "upon the impartial administration of justice depends the liberty of every individual, the sense which he has of his own security. [This is] in order to make every individual feel himself perfectly secure in the possession of every right which belongs to him."[99] In any political economy, principles of justice are expressed in particular laws that to differing degrees safeguard the equality of treatment, the enforcement of contracts and rights.[100] It is in this role that laws of justice are both conventional and "artificial" or utilitarian, though based on natural jurisprudence by which such laws can be evaluated and improved.

## The Problem of Equality

In a number of place in the *WN* Smith talks about equity, equality, or equality of treatment as ideals of justice.[101] He observes that individuals have by nature more or less equal natural talent. He explains, for example, that "the difference between the most dissimilar characters, between a philosopher and a common street porter, for example, seems to arise not so much from nature, as from habit, custom, and education."[102] Smith traces the distinctions among persons to differences in natural abilities, age, riches, and circumstances of birth. If each of us is more or less similar "by nature" and if each of us has certain basic natural rights, an ideal political economy would be one in which in an atmosphere of perfect liberty, natural jurisprudence is represented as closely as possible in laws of justice that create a system of equity. Moreover, Smith points out,

> No society can surely be flourishing and happy, of which the far greater part of the members are poor and miserable. It is but equity, besides, that they who feed, cloath and lodge the whole body of the people, should have such a share of the produce of their own labour as to be themselves tolerably well fed, cloathed and lodged.[103]

At the same time, Smith recognizes that the very existence of private property creates inequalities between those with and those without property. If in an ideal political economy one has the natural liberty to pursue one's economic interests and if civil government and laws of justice safeguard fair play and property, this will allow inequalities to multiply to such an extent that, as Smith admits, government must protect the rich against the poor. He sees those with wealth displaying greed and indolence, merchants and manufacturers often conspiring to pay low wages,[104] and landlords happily charging as much rent as possible.[105] What, then does he mean by equality and equality of treatment in the *WN*, particularly when he links equality to justice? Can a market economy, with its property inequalities, take care of the needy?

In a mature commercial society in which there is an atmosphere of "perfect liberty" and justice without restraints or favoritism on economic choices, "the whole of the advantages and disadvantages of the different employments of labour and stock must, in the same neighbourhood, be either perfectly equal or continually tending to equality."[106] Perfect liberty needs perfect competition, self-restraint, and cooperation as well, as we shall argue in the next chapter. Smith thought that in such an atmosphere, with an institutionalized framework of justice closely emulating natural jurisprudence, a competitive market would regulate the excesses made possible by the power of ownership. Competition would tend to curb these excesses, because pressures for capital reinvestment, at least most of the time, would prevent or disperse the power of wealth. Capital investment in productive enterprises also would eventually lead to full employment, thereby creating competition for workers and thus preventing employers from taking advantage of the laboring poor for very long. As we shall explain in more detail in Chapter 5, when labor productivity is a valued commodity, the "laboring poor" are better off in a commercial economy in which they at least have the opportunity to live where they please, choose their jobs, and bargain for better wages.

David Levy points out that according to Smith, it is economic growth, not property per se, that is the key to the advancement of a society. It is not the existence of private property that adjudicates economic inequalities, helps the needy, or improves the condition of the poor, for private property—as Smith acknowledges—perpetuates such needs and imbalances. But property is necessary for economic growth, and it is in this capacity that it improves economic opportunities. A system of natural liberty that reduces inconsistent regulations on merchants and manufacturers and frees laborers from their chattels to property owners and also protects property creates the conditions for such achievements.[107]

The issue is more complex than this, however. Smith argues that government is based on authority, authority that is derived from respect for wealth, intellect, rank, and distinction. At the same time he proposes an economy in which people engaged in commerce are at liberty to gain wealth. Does this newfound wealth become a threat to traditional authority? In the *LJ* Smith addresses this issue directly. Although he claims that most of us have little comprehension of the public interest, he nevertheless finds that "whenever commerce is introduced into any country, probity and punctuality always accompany it."[108] Men of commerce develop better manners if only to create an advantage for themselves in the marketplace. Thus commerce democratizes authority based on wealth, and what might appear to be a conflict between economic freedom and authority at least partially resolves itself.[109]

There are two other equalizing effects. When Smith talks of equity in the *WN*, he means that laws of justice should protect every person equally from the harms of other persons, and thus no laws that favor one group of persons should be tolerated. By equality Smith usually means "equality of treatment," that justice protects fair play. He does not imagine that any society will provide equal advantages for everyone, nor does he aspire to that form of egalitarianism. But he does believe that a political economy freed from restraints and regulations that exhibit favoritism and allow monopolies to flourish will be more likely to "tend toward equality" than will an atmosphere in which civil government tries to regulate such inequalities. One must take care to remember, however, that for Smith it is not merely the absence of regulations of the economy that will reduce inequalities, but nonregulation in the framework of a system of jurisprudence and in the context of competitive markets.

Finally, although benevolence is an imperfect virtue in that one has no perfect obligations to help the poor, one does have a duty (or society has a duty) to provide conditions that do not oblige the poor to remain in their abject condition. Smith is very clear in Book V that society has a duty to provide the means by which people can "improve their condition" or compete. Smith finds that public works, good public education, and an industrialized commercial economy with work opportunities can solve problems of poverty and inequality more satisfactorily than can other economic or political arrangements. More will be said about Smith's idea of a political economy in Chapters 4 and 6. Before turning to that topic, however, we must first see how natural liberty and natural jurisprudence, along with self-interest and the social interests, fit into the *WN*.

# Appendix

Kund Haakonssen laid out in schematic terms Smith's system of real rights[110]:

**Real rights**

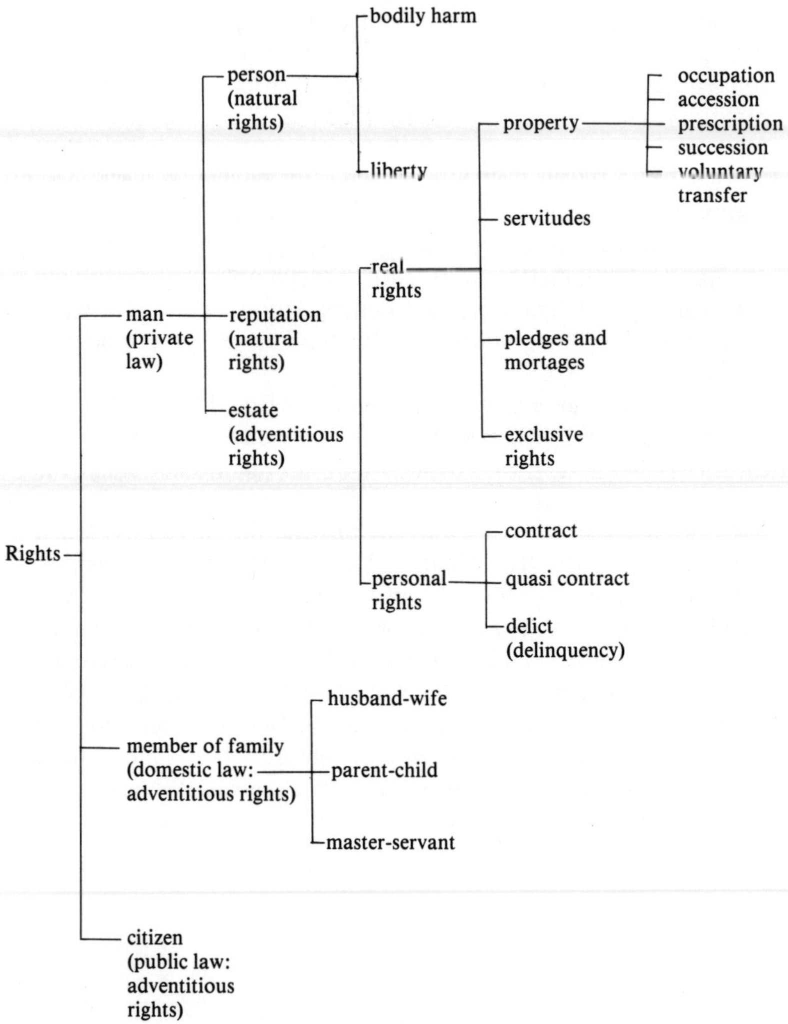

# 3

# Self-Interest, the Social Passions, and the Invisible Hand in the *Wealth of Nations*

> The *Wealth of Nations* is a stupendous palace erected upon the granite of self-interest.[1]

This statement, by the Nobel laureate and economist George Stigler, is representative of one popular interpretation of the *WN*. Stigler and others argue that self-interest, but not selfishness, is a dominant theme in the *WN*.[2] Exaggerated versions of this interpretation commit Adam Smith to a form of egoism in which selfishness is the "granite" of motivation, to a radical individualism, and/or to a glorification of the invisible hand as the final and impartial arbitrator of self-interests and as the nonhuman creator of economic good. Although no one person holds all of these views, a compendium of these readings of Smith has created a false impression of his economic and philosophical theses. Such readings, I shall argue in this chapter, come out of an overemphasis on self-interest or a misidentification of Smith's notion of self-interest as selfishness, a reading of "natural liberty" as a license to do as one pleases, an anthropomorphizing of the notion of the invisible hand, and/or a neglect of Smith's concept of justice. A number of critics have recognized the exaggerations of these interpretations,[3] and my task in this chapter will be to mount textual evidence that demonstrates the dubiousness of such readings.

I shall examine a number of seminal concepts in the *WN* in light of Smith's moral psychology developed in the *TMS* and his theory of rights developed in the *LJ*. One cannot deny the importance of the role of self-interest in the *WN*, but one needs to be reminded that it is Smith's notion, not that of Mandeville, Hobbes, or the later Social Darwinists. Individual self-interest is an important idea in the *WN*, but at best, it is only part of the "granite" underlying the *WN*, as the social passions and interests play equal roles in economic affairs. Although Smith defines "natural liberty" as a scenario in which "all systems of restraint being completely taken away," the "restraints being completely taken away" are those interfering with the natural liberty of free competition. As he writes in the *WN*, natural liberty has its own moral constraint, the constraint of justice. Economic order and economic well-being are achieved not merely because of an unregulated market. Rather, this ideal is realized only when economic actors perform with restrained self-interest, restrained by the desire and necessity to cooperate, by the laws of justice, and, most importantly, self-restrained by parsimony and prudence. Finally, I shall contend, the invisible hand is the outcome of market forces rather than the "ultimate governor which controls self-love,"[4] and so the "invisible hand" is only as impartial as are the conditions of restrained self-interest, liberty, cooperation, competition, and justice.

## Interests, Self-Interest, and Cooperation

Versions of the self-interest views that I outlined in the Introduction sometimes interpret self-interest in the *WN* as the atomistic, asocial, selfish pursuit of one's own economic interests. One well-known commentator has gone as far as to claim that "Adam Smith showed how the free market miraculously harnesses greed toward constructive ends,"[5] and Stigler finds that in the *WN* Smith did not carry his notion of self-interest far enough.[6] Yet in the first chapter I stated that whatever Smith's views are in the *WN*, he is unquestionably clear in his treatment of self-interest in the *TMS*. There self-interest is not identified as selfishness or greed and is derived from one, and only one, of the human passions. Smith avoids using the term *selfishness* except in reference to the selfish passions, to criticize Mandeville or Hobbes, or to describe self-interested actions that are greedy or unjust. Even so, the most selfish person both seeks the approval of others and sometimes "a desire of being what ought to be approved of," according to Smith. That is, even when we are acting purely in our own selfish interests, we are restrained by the desire for approval and the desire to meet the standards of what ought to be approved. Obviously, different individuals react with various degrees of concern for approval, but Smith is clear that the social passions coexist with, and are as powerful as, the selfish ones. Therefore, interest in others is as genuine as self-love is. Greed and avarice are vices, to be sure, but the virtue of self-interest, prudence, is on a par with benevolence.

This reading of self-interest in the *TMS*, I believe, is consistent with that in the *WN*. Self-interest in the *WN* has the following characteristics: First, it is usually equated with the natural desire to better one's condition, plus the natural liberty to look after one's own welfare. Second, however, self-interest is not identified with selfishness or, worse, with greed. Rather, Smith recognizes the virtue of prudence in parsimony and in the economy of one's economic desires. Third, self-interest is both driven and restrained by the desire for approval. Connected with the desire for approval, fourth, economic self-interest makes sense only in the atmosphere of mutual cooperation, a phenomenon that traces its origins to the social passions and interests. These passions and interests have as much weight as do the selfish passions in economic affairs. In the *WN*, then, self-interest is best defined not as selfishness but, rather, as personal interest, "only the interests of the individual concerned in the matters with which he is most intimately concerned."[7]

According to the *TMS*, "every man is certainly, in every respect, fitter and abler to take care of himself than of any other person."[8] In the *WN* this translates into the natural liberty to take care of one's own economic interests without external restraints. "The principle which prompts to save, is the desire of bettering our condition, a desire which, though generally calm and dispassionate, comes with us from the womb."[9] Smith recognizes that economic interests are more closely identified with the natural passion of self-preservation and the natural liberty to take care of oneself than with benevolence. But this does not imply, in itself, that in exercising one's self-interests one is, or can be, merely selfish or asocial, that pursuing one's self-interests is always greedy or bad, or that such pursuits exclude the social passions and interests.

In the context of the natural desire for self-preservation and concern for bettering one's condition, that individuals are not benevolent is irrelevant. Indeed, one does not expect benevolence in economic situations.

> Man has almost constant occasion for the help of his bretheren, and it is in vain for him to expect it from their benevolence only. He will be more likely to prevail if he can interest their self-love in his favour, and shew them that it is for their own advantage to do for him what he requires of them.[10]

Moreover, "nobody but a beggar chuses to depend chiefly upon the benevolence of his fellow-citizens."[11] Thus receiving charity is not an honorable economic expectation, nor should it be, Smith contends.

Given Smith's questions about the role of benevolence in economic exchanges, another way to think about what he means by self-interest in the *WN* is to borrow from contemporary rational choice theory. He could be saying that at least in economic affairs, "persons [should] be conceived as not taking an interest in the interests of those with whom they exchange."[12] This is called *non-tuism*,[13] which is different from straightforward psychological egoism. Non-tuism is not identified with self-interest or greed but, rather, with a disinterest in the concerns of other economic actors as they engage in the exchange process. It implies that in economic exchanges there is mutual unconcern in others' interests but not that each is motivated only selfishly or even by self-interest in every context.

The fact that benevolence is neither connected with economic motivation nor is a virtue in economic relationships does not imply that the *TMS* and the *WN* arrive at contrary conclusions, nor does this fact negate the role of the social passions in economic activities. According to Russell Nieli, the reason that benevolence is allegedly neglected in the *WN* is because it is a virtue of intimate relationships. The objects of our benevolent affections are primarily those closest to us: our family, relatives, and friends. Economic relationships, on the other hand, are material, disinterested, and not intimate. Therefore the virtue of benevolence would be out of place in such relationships.[14]

There is another, more important reason that benevolence does not, and should not, play a role in economic affairs. Recall that although benevolence is a virtue, it is voluntary and nonenforceable. One has only imperfect duties to charity, and one need not be benevolent in order to be virtuous or, more specifically, in order to be fair. On the other hand, as Smith argues in the *Lectures of Jurisprudence*, one has perfect duties not to harm another or to interfere with that person's own liberty. Those duties are universally applicable or enforceable. Therefore justice, unlike the other virtues, can be codified into law, and one can specify the rules of fair play. Benevolent or malevolent economic actions follow only vaguely delineated general rules, and they cannot be specified in law without placing undue burdens on some. But fair play, as Smith defines it, as a negative injunction is specifiable, clear-cut, and enforceable. Thus it is justice, not benevolence, that is crucial to economic exchanges, because rules for fair exchanges that are not harmful to others can be enunciated without bias toward any individual or groups of individuals. Indeed, far from excluding the social passions and interests in economic affairs, the presence of justice reinforces these passions and interests as essential features of a political economy, because justice is the virtue of the impartial social passions and interests. In a political economy, economic exchanges based on rules of justice rather than on benevolence are thus the ideal, an ideal that further undermines the claim that self-interest is the "granite" of the *WN*.

Because benevolence is not given a role in the *WN*, Max Lerner—in his introduction to the Modern Library edition of the *WN*—went so far as to claim that Smith "gave new dignity to greed and a new sanctification to the predatory impulses . . . he rationalized the economic interests of the class that was coming to power."[15] A. O. Hirschman declared that "the main impact of *The Wealth of Nations* was to establish a powerful *economic* justification for the untrammeled pursuit of individual self-interest."[16] But this is not what Smith had in mind. In the *WN* he clearly distinguishes between acting in one's self-interest and greed. He condemns the moral decay of the rich who do not invest their capital and argues that high profits destroy parsimony and lead to a reduction in capital. High profits can also be a disadvantage to the public interests, and avarice prevents good economic performance.[17] Thus, although benevolence may not be a consideration in economic exchanges, greed is antithetical to a well-run system and so should be discouraged or prevented. Moreover, according to Smith, self-interest is or should be restrained by parsimony, a sign of the virtue of prudence in the *WN*: "Every prodigal appears to be a publick enemy, and every frugal man a publick benefactor."[18] "Though the principles of common prudence do not always govern the conduct of every individual, they always influence that of the majority of every class or order."[19]

The bracketing of benevolence in economic exchanges does not imply the dismissal of a role for the social passions. At least part of our economic motivation is not merely to better our condition but also to receive the approval or admiration of others. According to Smith, it is not merely self-interest that motivates us but, in addition, a desire for respect. We respect and admire economic well-being; that is, we share a fellow feeling with those who have wealth. Therefore, part of one's competitive drive for economic well-being stems from the desire for the approval of others, and this desire, according to the *TMS*, comes from the social passions. Smith is careful, however, to notice that part of that desire is also envy of another's good fortune, for he recognizes that we tend to admire the rich and avoid the poor.[20] So in the *WN* the desire for approval appropriates the selfish, the unsocial, and the social passions.

In addition to the desire for approval and the importance of fairness in economic affairs, the social passions play another, more important role in the *WN*. To defend the thesis that self-interest, although not necessarily selfishness, is the dominant motivating force in the *WN*, one usually cites Smith's famous remark that

> it is not from the benevolence of the butcher, the brewer, or the baker, that we expect our dinner, but from their regard to their own interest. We address ourselves, not to their humanity but to their self-love, and never talk to them of our own necessities but of their advantages.[21]

Smith deliberately uses the "butcher, the brewer, and the baker" example to illustrate that although our self-interests appear to dominate in the economy, cooperation is both natural and essential. None of these tradespeople acts benevolently, yet each depends on our approval for his or her self-interest. It is to everyone's mutual advantage to produce good bread, beer, and meat, and these tradespeople will fail economically if they do not take this into account. Moreover, in the small-town atmosphere implied in the example, these tradespeople depend on cooperation to stay in business. I suggest that it is the natural interest in trading with others that triggers such cooperation. It is also natural in such an atmosphere to desire the approval of others, for example, for the quality of one's bread, and at the same time it is a self-interested economic necessity. Although the butcher and baker are not necessarily benevolent, they not only are not malevolent, in the "corner grocery store" economy Smith depicts, these shopkeepers cannot function without cooperation. The fact that "we . . . never talk to them of our own necessities but of their advantages" shows that some sort of fellow understanding underpins economic relationships and, in particular, competition. Economic interchanges even between self-interested parties are not purely adversary, according to Smith, because they depend on cooperation and coordination. This assertion led the economist P. L. Danner to contend that exchangeable value, the source of wealth and economic growth, according to Smith in the *WN*, arises from the phenomenon of sympathy, because, according to Danner, it is through sympathy that we become interested in the fortunes of others and thus enter into mutual and reciprocal agreements.[22]

I argue that mutual coordination and cooperation, though essential to economic exchanges, are derived from the social passions and interests. But Smith is careful not to use the term *sympathy*, as it is the social passions that prompt our interest in, but not necessarily our approval of, others. In the *TMS* Smith distinguishes social and selfish passions and their interrelationships in a fairly abstract way. In the *WN* he shows how these kinds of two passions and the interests and motivations that spring from them, far from distinct, are interrelated.

Smith's analysis of the division of labor in the *WN* is an illustration of the intersection of self-interest and social interests. "This division of labor . . . is the necessary, though very slow and gradual consequence of a certain propensity in human nature . . . to truck barter, and exchange one thing for another."[23] This propensity prompts us to cooperate with others, and it is also to our mutual advantage to do so. Our natural desire to cooperate motivates us to barter with other individuals while at the same time appealing to their self-interest in the exchange. It is also to everyone's personal advantage to enter into such a system, because by focusing on his or her skills, each person can increase productivity and exchange his or her specialized labor for other goods. As Smith explains in this regard: "The gains of both are mutual and reciprocal, and the division of labour is in this, as in all other cases, advantageous to all the different persons employed in the various occupations in which it is subdivided."[24]

In other places in the *WN*, too, Smith notices that economic exchanges cannot operate in the "vacuum" of self-interest but require cooperation and coordination. In distinguishing four uses of productive capital—production, manufacture, transportation, and selling—Smith writes: "Each of those four methods of employing a capital is essentially necessary either to the existence or extension of the other three, or to the general conveniency of the society."[25] He also claims that free trade requires the cooperation of two parties or two countries.[26] Tradespeople often work together, sometimes even in collusion, despite their competitive relationships.[27] This cooperation is also seen in the relationships between townspeople and farmers, according to Smith. Cities provide the country with trade and manufactured items, and the farmers and country provide the cities with food and raw materials. Both gain from such natural, reciprocal, and mutually advantageous arrangements.[28]

In all these case, cooperation is both natural and required for the advantage of our self-interests. Thus our individual and social interests are reconciled through cooperation, the division of labor, barter, and trade. Although non-tuism appears to dominate in economic transactions, self-interest is not the only motivation, and cooperation is both natural and essential. Thus cooperation is identified neither with enlightened self-interest nor with benevolence but is the expression of the social passions in economic affairs.

The question is whether we participate in cooperative ventures because it is in our interest to do so, or whether economic exchanges, which require close cooperation, arise out of our social passions and satisfy the selfish passions as well. Hobbes allegedly favored the first conclusion; in the *TMS* Smith might have opted for the second. In the *WN* he finds these two sets of passions and interests more closely interrelated. It is both a natural passion and an advantage to cooperate in the division of labor, and one interest does not derive from the other. We are naturally both self-interested and cooperative, but these interests are seldom distinct, and we do not ordinarily sort out our motivations accordingly.

In the *WN* self-interest is defined neither as selfishness nor as greed; rather, it pertains to both self-love and other interests of personal concern, interests having to do not merely with bettering our condition or desiring approval but also with intimate social interests. Indeed, in the *WN*, Smith equivocates in his use of the term *self-interest*. The term is usually used in the context of talking about individual economic exchanges, exchanges that directly affect the well-being of each economic actor. In such exchanges our interests are literally at stake because it is a self-initiating action that concerns our future well-being. At the same time "self-interest" may refer to other private or personal interests as well. Smith makes a further distinction between private interests and public interests: Private interests are interests of individuals or groups of individuals that do not concern themselves with society as a whole or the public good. Public interests are those interests that concern society and every member of society equally.[29] These distinctions are important to the discussion of the efficiency and fairness of the market.

Implicit in the *WN* is the assumption that human beings are social beings who are interested in the fortune of others, at least to the extent that self-interests often overlap with the interests of others and to the extent of cooperating and desiring approval. So even though economic actors engage in business for their own interests, the social passions—translated into cooperation and the dependence of each of us on others to develop our interests—regulate behavior, the degree of regulation depending on the extent to which each individual internalizes norms of societal approval and/or is dependent on the cooperation of others. Smith would grant that not every individual is equally concerned about what others think, nor is each person equally cooperative; some are selfish, others are totally unconcerned; still others are unselfish and even self-sacrificing. But Smith would also argue that no individual is not without interests in others, nor can any of us fail to take into account the social pressures of approval or the necessity, if not the desire, to exchange, coordinate, and cooperate with others.

One might also contend that I have exaggerated the role of the social passions and interests and thus the role of cooperation in the *WN* and so played down the dominating role of self-interest. It appears that self-interest motivates the desire to complete and to improve one's economic condition. But the desire to engage in exchanges, the desire for approval, and the desire for praiseworthy conduct come from the social passions, and as I believe, for Smith the importance of justice in the economy cannot be exaggerated. One cannot underestimate the social passions and interests, for they trigger one's interest in engaging in exchanges, and without cooperation no economy could function. Moreover, cooperation is not equated with benevolence, nor is it identified with justice. So one can engage in cooperative ventures that are neither benevolent or fair. Therefore, although I disagree with Danner concerning the role of sympathy in the *WN*, I do agree that self-interest makes sense only in a framework of economic exchanges that require cooperation and coordination. Social passions and interests are of central importance in the *WN*, and self-interest is neither the only motivating force nor the "granite." The notion of self-interest in the *WN* is that of the *TMS*.

## Sympathy in the *Wealth of Nations*

In the foregoing I have assiduously avoided using the term *sympathy* because of its blatant absence in the *WN*, and I have argued that the social passions and interests are the basis for cooperation. The mere fact that the notion of sympathy is not developed in the *WN* is not enough evidence to conclude that Smith dropped that term from his moral psychology, because he retains it as a prominent concept in the last edition of the *TMS*. Therefore, although it is not as clear that sympathy plays as prominent a role in economic relationships as in morality, it is safe to say that Smith presupposed that concept in the *WN*. But in its technical use, the term is not employed in that text. Some commentators, finding that sympathy is not mentioned in the *WN*, conclude that Smith thought that sympathy had nothing to do with economic affairs.[30] In contrast, other interpreters, trying to find parallels between the *TMS* and the *WN*, have overstated the role of sympathy in the *WN*.

Marjorie Clay suggests that self-interest is the dominating principle in both the *TMS* and the *WN*. She then claims that if one can make a case that self-interest in both works has a similar meaning—a point that I just defended in the last section—and if one can derive sympathy from self-interest, then one can demonstrate a marked congruity between the two works.[31] This is an interesting thesis, because it in part solves some of the apparent discrepancies between the two texts. But such a thesis is virtually impossible to establish. Sympathy and self-interest are different kinds of phenomena. Self-interest is a motivating force. Sympathy is not a passion and is not derived from a passion but, rather, is the means through which we understand (but do not feel) the passions of others and ourselves. Therefore, sympathy has no role in motivation. Sympathy is connected with the social passions, because our interest in others provides the motivation to understand what another person is feeling. Sympathy is indirectly linked to self-interest because it might be in our self-interest to understand the passions of others, and understanding the passions of others might influence our self-interests. But, as we saw in Chapter 1, Smith makes clear in the *TMS* that sympathy entails a genuine fellow feeling not connected with selfishness or self-interest or, indeed, with any passions or interests at all. Thus he would not want to derive sympathy from self-interest, and such a move would be contrary to his criticism of Hobbes for making a similar suggestion.

Opposing the idea that self-interest is the dominating theme in the *WN*, P. L. Danner claims:

> In the *Wealth* the concept, which is designed to match in the world of economic relations what sympathy was supposed to do in personal relations, is . . . *exchangeable value* . For Smith conceived of the economic good in very concrete terms as something which would excite individual self-interest and yet demand from its pursuants that they cooperate and share. At the same time he founds the struggle for economic well-being on the aspiration for sympathy.[32]

Danner is not saying that sympathy and exchangeable value are identical terms. Sympathy is a psychological phenomenon that "interests us in the fortunes of others" and allows us to put ourselves in another's situation. Exchangeable value is a measure of productivity and real wealth. These are different sorts of notions. It also is, at best, unclear that Smith intentionally connected the function of sympathy in the *TMS* and exchangeable value in the *WN*. Danner, however, is making another point. What underlies exchangeable value and enables economic exchanges, productivity, and wealth has roots in Smith's earlier notion of sympathy. There are four reasons for this, according to Danner. First and most trivially, we respect and admire economic well-being; that is, we share a fellow feeling with those who have wealth. Therefore, part of our competitive drive for economic well-being stems from our desire for the approval of others. Second, Danner states, the "it is not from the benevolence of the butcher" example demonstrates sympathy as well as self-interest at work. The fact that "we . . . never talk to them of our own necessities but of their advantages" shows that fellow understanding underpins economic relationships and, in particular, competition. Third, as Danner correctly recognizes, economic interchanges even between self-interested parties are not purely adversary, according to Smith, because they depend on "mutuality, reciprocity, and coordination."[33] Danner contends that the latter, in turn, arise from the phenomenon of sympathy, because it is through sympathy that we become interested in the fortunes of others and thus enter into mutual and reciprocal agreements. Fourth, exchangeable value—the sources of wealth and economic growth, according to Smith—based as it is on productivity as well as capital and rent, is labor dependent. For, as Smith says, "labour, therefore, it appears evidently, is the only universal, as well as the only accurate measure of value, or the only standard by which we can compare the values of different commodities at all times and at all places."[34] Labor and the division of labor are dependent on cooperation and coordination, and, Danner claims, cooperation and coordination develop out of sympathy.

Danner explained an important and often-neglected point: Smith's recognition that allegedly self-interested competitive economic interchanges depend on an economically harmonious system of coordination and cooperation, a phenomenon about which a great deal was said in the preceding section. I question tracing this harmony to sympathy, however. According to Danner's argument, mutuality, reciprocity, and coordination make possible the development of economic self-interests. Thus it is in our self-interest to be sympathetic. Is sympathy, then, derived from self-interest? Is Smith's alleged individualism based on the priority of self-interest? In the last paragraphs I argued that Smith did not intend to derive sympathy from self-interest, as they are distinct kinds of phenomena. Moreover, in the *TMS*, although approval derives from sympathy, the desire for approval comes from the social passions. Nor is it clear from either the *TMS* or the text of the *WN* that mutuality, reciprocity, and coordination arise out of a fellow feeling with, or understanding of, the passions of our economic competitors. Rather, as I have contended, mutuality, reciprocity, and coordination come from the social passions and interests and are of equal status with self-interest. Therefore, exchangeable value, though a central term in the *WN*, derives from the social interests, not from sympathy specifically.

Because of the importance of sympathy in the *TMS* and because sympathy is a source of approval, Robert Lamb stated that in both the *TMS* and the *WN*,

> it is the individual's desire for the approval of other men which motivates him to virtuous acts and to property accumulation. . . . [T]heir fundamental desire is to receive sympathy from other men for their material as well as their moral situation. . . . Therefore "sympathy" rather than self-interest is the basis of property in Smith's system.[35]

Interestingly, as we saw earlier, Smith seldom mentions the term *property* in the *WN*. In the *WN* it appears that he accepts his own arguments from the *LJ* in defense of private property. For example, in the *WN* he talks about the "sacred rights of private property,"[36] although in the *LJ* he says that the right to property is an adventitious, not a natural, right. It is clear that the notion of private property is a basic presupposition of Smith's political economy. But is it connected with the desire for approval and with sympathy?

In the *TMS* Smith asserts that at least part of the motivation to accumulate property and capital is the desire for the approval of others, although this approval does not always have a basis in virtue.[37] In the *WN*, part of the motive for property and capital accumulation is the desire for approval or respect, as Smith recognizes that we tend to admire the rich and avoid the poor. But as I argued earlier, an equally important motive is self-interest linked to a desire to improve our condition. Both motives are operative, but we can scarcely conclude, based on the text, that the desire for approval takes precedence. Moreover, however much we wish to link the *TMS* to the *WN*, sympathy is a source of approval but not the source for the desire for approval. In the *WN* the desire to better our condition, the desire for approval, and our natural interest in others are the primary motives for accumulating property, remembering, of course, that self-interest is not identified as selfishness, and the desire for approval is not the same as benevolence. But because sympathy is not a passion and does not function as the desire for approval, sympathy cannot be the basis for property in the *WN*. Thus, although Smith does not reject the notion of sympathy in the *WN*, it is merely assumed and does not play an active role in his theory of political economy.

## Self-Command

Because Smith added the notion of self-command to the sixth edition of the *TMS*, after he had completed the *WN*, at least one commentator has suggested that

> the ethics of self-command is the culmination of Smith's *Theory of Moral Sentiments*; it is also the foundation of his Jurisprudence and political economy. Freedom, both moral and economic, meant to him self-reliance, the ability of the individual (through his moral sentiments) to "command" himself according to the objective principles of equity, natural law, prudence and justice.[38]

Unfortunately for this reading of Smith, self-command is not mentioned in the *WN*. It is unclear, then, whether Smith meant to link that notion to the ideal economic actor in the *WN*. But in the *WN* he does set out the qualifications for such a person. The ideal economic actor, endowed with the desire to better his or her condition and the ability to care for his or her own interests better than others can care for them, naturally engages in cooperation and in the care of his or her interests and exercises natural liberty within the guidelines of fair play.[39] The ideal economic person is also self-restrained by parsimony rather than greed. Such a person, then, is the economically prudent person in command of himself or herself.

But self-command, by itself, is not enough to serve as the foundation of political economy. There are two reasons for this. According to Smith, despite his emphasis on the liberty and self-reliance of individuals, a society needs institutionalized laws of justice. He has little faith that individuals will exercise self-restraint equally. Second, as I shall argue in the next section, cooperation, though equal to self-interest as a motivating principle in the economy, does not guarantee fair play. So although one can compare an ideal economic actor with the person in command or with himself or herself, Smith does not make that connection explicitly, and the ideal economic actor alone cannot achieve in the marketplace what internally he or she can achieve as a virtuous person.

## The Invisible Hand

The term *invisible hand*, for which Adam Smith and the *WN* are most famous, appears exactly once in that work; it also appears only once in the *TMS*. In the latter context the term refers to the unwitting distribution by the rich to the poor:

> The rich only select from the heap what is most precious and agreeable. They consume little more than the poor, and in spite of their natural selfishness and rapacity, though they mean only their own conveniency, though the sole end which they propose from the labours of all the thousands whom they employ, be the gratification of their own vain and insatiable desires, they divide with the poor the produce of all their improvements. They are led by an invisible hand to make nearly the same distribution of the necessities of life, which would have been made, had the earth been divided into equal portions among all its inhabitants, and thus without intending it, without knowing it, advance the interest of the society, and afford means to the multiplication of the species.[40]

Because the rich cannot consume what they demand and purchase, there is a large residue of goods that the rich are "obliged to distribute" to the rest of us. Moreover, because the *TMS* sees the natural harmonious order of society as in some way reflecting God's order, it is sometimes suggested that the invisible hand in the *TMS* is meant to be God's hand in the economy. This is probably an incorrect reading, because the context in which the expression appears in the *TMS* implies, as in the *WN*, that the invisible hand is the "fallout" from self-interested and sometimes selfish competitive economic actions. Nevertheless, because of Smith's eighteenth-century religious background and upbringing, it is not surprising that he would use a term that has religious connotations to describe an economic phenomenon. If the ideal economic order, governed by the invisible hand, reflects closely the natural order, an order in some way or other commanded by God, Smith's use of the term invisible hand to refer to economic activities is consistent with this sort of metaphysical speculation.

In the *WN* the "trickle-down" analysis of the invisible hand as the unwitting distributor of leftover goods is dropped. There the invisible hand is an explanatory term to explain why it is that economic exchanges unintentionally create economic growth and well-being even for those not directly participating in the exchanges. In Book IV Smith states:

> He generally, indeed, neither intends to promote the publick interest, nor knows how much he is promoting it. By preferring the support of domestick to that of foreign industry, he intends only his own security; and by directing that industry in such a manner as its produce may be of the greatest value, he intends only his own gain, and he is in this, as in many other cases, led by an invisible hand to promote an end which was no part of his intention. Nor is it always the worse for the society that it was no part of it. By pursuing his own interest he frequently promotes that of the society more effectually than when he really intends to promote it.[41]

Even though Smith mentions the invisible hand only once in the *WN*, this notion has been the subject of volumes of commentary, interpretation, and extrapolation. The invisible hand has been interpreted as the free market created out of perfect competition among self-interested individual economic actors, a market that functions as an independent impartial arbitrator and stabilizer for the economy. For example, a well-known commentator on Smith, A. L. Macfie, interprets the invisible hand to be "the ultimate governor which controls the self-love of individuals . . . the ultimate natural harmony of individual interests."[42] Robert Lamb calls the invisible hand "the universal reign of the impartial spectator, the invisible but rational hand, which rational producers for their self-interest must obey."[43] Let us look at the *WN* to see whether these sorts of readings of the invisible hand are viable.

Macfie argues that in the *TMS*, the impartial spectator, along with general (moral) rules, an internal sense of fair play, and the virtue of prudence serve as "controls" for human behavior. However, according to Macfie, in the *WN*, prudence is replaced by the desire to better one's condition, and there is no mention of the impartial spectator or general rules. The natural social order is replaced by the "natural harmony of the economic order," but there is no guarantee that economic actors will practice internal self-restraint in that economic order. Therefore, Macfie reasons, in the *WN* the invisible hand replaces the impartial spectator as controller of the economic order, the "ultimate governor." Macfie is not suggesting that the invisible hand is the impartial spectator but that in the *WN* it functions as a regulator of economic behavior, just as the impartial spectator functions as a regulator of moral behavior in the *TMS*.

As I stated earlier, however, in the *WN* Smith does not replace the prudent person with the desire to better one's condition, as Macfie maintains, and Smith talks in both texts about the prudent person and bettering one's condition. Moreover, there are some crucial differences between the impartial spectator and the invisible hand. The impartial spectator, however objective, is a prototype of, and functions only as, an embodied human being. But as a human being, the impartial spectator can never be perfectly impartial or absolutely disinterested. Judgments of the impartial spectator are, at least partly, contextual; that is, they are based on what a particular society approves of in the kind of situation embodied in that context. These judgments are deliberately and self-consciously made by an individual disengaging himself or herself into the role of the spectator. Macfie is correct in saying that in the *WN* Smith has less faith in the ability of economic actors to practice self-restraint. But in the *WN* this role of personal impartial judgment making is replaced not by the invisible hand but by the laws of justice.

The invisible hand, on the contrary, is simply the result of economic interchanges in free-market operations. Although the invisible hand has been interpreted to be the market as it acts impartially to regulate economic behavior, its so-called impartiality is a result of competition, competition among economic actors that, if unregulated and fair, creates lower prices, jobs, more goods, and economic growth, results that are ordinarily not always deliberately intended by the economic actors. As an alleged "actor," in fact, the invisible hand is merely a result of economic exchanges and cannot function without them. It is not impartial in the sense that it is an autonomous "hand" that steps back, grasps the economic situation, and dispassionately decides how economic behavior will be balanced and restrained. Because the invisible hand is created out of competition among individuals, it arbitrates the market fairly only insofar as there is a balance of more or less equal competition among economic actors. That it controls self-interests is possible only to the extent that there exists a balance of those self-interests in the marketplace, and not the converse. Therefore, the invisible hand is the side effect of free-market exchanges, not the self-conscious, deliberate "ultimate governor" of economic behavior or the creator of economic good. The invisible hand is an explanatory term to describe how self-interested behavior can produce unintentioned economic well-being.[44] In support of this interpretation Smith says in the *TMS*

> The man of system . . . seems to imagine that he can arrange the different members of society with as much ease as the hand arranges the different pieces upon a chess-board. He does not consider that the pieces upon the chess-board have no other principle of motion besides that which the hand impresses on them; but that, in the great chess-board of human society, every piece has a principle of motion of its own, altogether different from that which the legislature might chuse to impress upon it.[45]

The comparison between the impartial spectator and the invisible hand is not completely satisfactory, because such a comparison personifies the latter. Economic good may be produced as a result of competition, competition that is self-regulatory if it is balanced. But no process deliberately makes moral judgments about economic good or regulates self-interests "which rational producers for their own self-interest must obey," because there is no "entity" similar to the personification of the impartial spectator to initiate and regulate this process.

If the invisible hand is merely the outcome of the sum of economic behaviors in a free market, it appears to be an explanatory term referring to the functions of the "economic machine" and the consequences of such functions. Because the invisible hand is itself not an actor, how it works and how well it works depend on a number of factors. The factors that create and "drive" the invisible hand include the following: According to Smith, first and most obviously, the free market depends on the condition of natural liberty, the absence of constraints on market activities. And about natural liberty Smith writes: "Every man, as long as he does not violate the laws of justice, is left perfectly free to pursue his own interest his own way, and to bring both his industry and capital into competition with those of any other man, or order of men."[46] So second, natural liberty should function only within the framework of justice that protects citizens from harms and guarantees fair play. What is interesting in Smith's analysis is that we accept a framework of a system of justice not merely because of the authority of the law or the utility of the security of our interests. Justice is the impartial virtue of the social passions. Laws of justice that impartially apply to everyone equally, enforcing fair play, are acceptable even when they are not in our self-interest or in the private interests of a group or association. Our objection to laws, according to Smith, is an objection to those regulations that favor or punish some people or some groups of people. Thus it is unfair laws, not regulation per se, that Smith questions.[47]

Third, perfect liberty must be accompanied by self-restraint, an internal sense of prudence. Although economic activities can be non-tuistic, selfish or greedy economic activities can damage competition and hurt the free market. Smith recognizes this when he criticizes dealers for narrowing competition in their own interest,[48] employers who contrive to pay low wages,[49] or merchants and manufacturers who advocate trade restrictions.[50] Monopolies are usually a product of such greed, and "it is thus that the single advantage which the monopoly procures to a single order of men is in many different ways hurtful to the general interest of the country."[51]

Fourth and less obviously, a free market works most efficiently when there is competition between more or less equal parties in which one party does not have undue advantages not available to the other. Again and again in the *WN* Smith extols the virtues of competition as being to the advantage of the producer, the laborer, and the consumer.[52] For Smith the greatest cause of inequality or unfairness is any restrictive policy that deliberately gives privileges to certain kinds of businesses, trades, or professions. Yet he also believes that ideally, competition should be among parties of similar advantage. A system of perfect liberty, he argues, should create a situation in which "the whole of the advantages and disadvantages of the different employments of labour and stock . . . be either perfectly equal or continually tending to equality."[53] Smith sees perfect liberty as a necessary but not a sufficient condition for competition, but perfect competition occurs only when both parties in the exchange are on more or less equal grounds, whether it be competition for labor, jobs, consumers, or capital. This is not to imply that Smith favors equality of outcomes. Clearly he does not, nor does he think that such equality will be a result of a free-market economy.[54] But the market is most efficient and most fair when there is competition among similarly matched parties.

Fifth, as I argued earlier, no market can function without coordination or cooperation. Smith shows how unhonored contracts,[55] unfair banking practices,[56] and price conspiracies[57] violate market coordination. Thus the degree of harmony and coordination affects whether or not a market functions and functions well.

There is a problem with this claim, however. If economic actors are naturally cooperative as well as competitive, why do we need laws of justice? Why is the spirit of cooperation within the framework of competition not enough to balance self-interests and guide the market? With this sort of point in mind, Smith has been attributed to be the advocate of a form of "benevolent capitalism" as the ideal. Self-restrained economic actors in cooperation with one another can, by themselves, create an ideal economy beneficial to everyone.[58] But there are a number of problems with this conclusion. First, it is not benevolence that is desirable in an ideal economy, but fairness. That is, a benevolent person can be unjust, and benevolence is not enforceable. Second, cooperation and coordination by themselves do not guarantee fair play. Often people act in collusion with one another in the economy. Third, in general, according to Smith, people are shortsighted and do not always act rationally. Lack of foresight, our personal situation, and ingrained habits often lead to our placing our own interests and those of our friends and colleagues ahead of the public interest. Even then, because we are neither perfectly rational nor foresighted, we often neglect our own long-term private interests.[59] This is why Smith states: "I have never known much good done by those who affected to trade for the publick good."[60] Few of us are capable of understanding what the public good is or how it can best be served, and we often confuse benevolence or charity with fairness and impartiality. This is also why Smith objects to most government regulations—not only are they likely to give advantages to one group over another, but also public servants who administer such regulations are themselves not always impartial or have the proper disinterested outlook. On the other hand, laws of justice that embody principles of commutative justice protect interests equally. Therefore laws of justice act as an impartial spectator to safeguard public interests and restrain cooperative as well as competitive market ventures. Keeping in mind Smith's distinction between private and public interests, it is these latter interests that are not always taken into account in economic exchanges and so must be guarded by a system of justice. Justice, then, safeguards both fair play in market exchanges and the public welfare.

The invisible hand, far from being the impartial spectator of the market, can operate only with the self-interested and cooperative actions of economic actors, and it is efficient only to the extent that it exists within the context of perfect liberty, coordination or economic harmony, equally advantageous competition, and fair play. It is not true, as one commentator suggests, that "the market cares not for fairness, but only for efficiency,"[61] as the market cannot be optimally efficient and competitive except in the context of fair play. Far from being the "ultimate governor" or the "universal reign of the impartial spectator . . . which rational producers for their self-interest must obey," the invisible hand can "regulate" only insofar as the five conditions for a perfectly free market are met.

The perfectly competitive market, then, far from being "free from morality," is a result of a number of conditions, each of which has a moral dimension. James Buchanan characterizes the perfectly competitive market as an "exemplar of rational morality, rather than a 'morally free zone'."[62] But this, too, is perhaps an overexaggeration. Like the description of the invisible hand as the impartial spectator, the idea that the market is an exemplar of rationality implies not that the market is an outcome of rational behavior but, rather, that the market itself can act rationally. Given the five conditions, Smith does conclude that a free competitive market can unintentionally promote economic well-being as well as economic growth. Simply put, competition creates lower prices, higher demand, more employment and economic growth, and, eventually, higher wages and full employment. But these outcomes are unintentioned outcomes of the competitive market and not a result of its rational intentionality.

Such a system, if fully operative, avoids restrictive economic politics, abolishes trade protections, and functions without special privileges that are advantageous only to some. Most importantly, the ideal of a free competitive market is closer to what Smith conceived of as the natural order. In such a system, economic exchanges are made freely by individual actions on the basis of restrained self-interest, cooperation, fair play, and the laws of negative justice.

## Conclusion

It is evident from the foregoing analysis that the *TMS* and the *WN* are not contradictory works. Adam Smith is consistent in his use of the term *self-interest* throughout the two texts. In both he defines self-interest as self-love rather than selfishness or greed. In the *TMS* Smith argues that self-interest is not evil and that prudence is a virtue equivalent to benevolence and justice. Similarly, in the *WN*, self-interested economic behavior is not an evil, and self-restrained economic behavior can be a good. Prudent, self-restrained economic activities contribute to fair competition and economic well-being in a free market, the extent of whose "providence" depends in great measure on the behavior of the economic actors.

The social passions and interests also play an important role in both texts. Arguing against Mandeville, Hobbes, and Hutcheson, in the *TMS* Smith places the social passions on a par with the selfish ones. In the *WN* the social passions reappear intertwined with the selfish passions in the form of our natural desire to truck, to barter and exchange, and our wish to be admired or approved of. Thus we naturally engage in economic activities and cooperate in economic ventures while at the same time satisfying our competitive self-interests in such exchanges, and it is this cooperation that makes competition possible. Therefore, self-interest is not the "granite" of the *WN*.

Because self-interest is not identified with selfishness or greed and because of the social passions and interests, economic actors are not merely self-interested rational utility maximizers. They are naturally cooperative, and they both seek and need the approval of others. But might it not be the case—even though ordinarily we are both selfish and cooperative—that in our rational moments our economic preferences are self-interested? Smith's answer would be both yes and no. He notices that economic actors, either alone or in collusion with others, often act in a self-interested or a group-interested manner. Such activities are sometimes restrained, but at other times they create monopolies, interfere with free trade, skew competition, and thus disrupt a free market. So a rational person or a group of persons cannot merely act in his or her or their own narrow interests without threatening the viability of those private interests. But for Smith it is not merely in our long-term enlightened self-interest to be parsimonious, cooperative, and just. It is also part of our nature to be so, even though we do not act accordingly a great deal of the time. For Smith, then, a rational person is prudent, cooperative, and fair both naturally because of the social passions and interests and because it is in his or her self-interest to be so.

The invisible hand is both more simple and more complex than Macfie envisions. The free market is not an autonomous overseer of self-interest; at the same time how it functions and whether it functions well depend on a number of complicated variables. The market is not merely driven by, or functions only because of, self-interested competition. Conversely, market failure is due to a number of factors, not merely a failure of self-interest or competition. The market is not self-governing, nor can it increase economic well-being unless all factors affecting it are operating optimally.

The invisible hand, then, is a dependent, not an independent, variable in economic activities.

The necessity of justice is a third theme running through the *TMS*, the *LJ*, and the *WN*. Justice, as an internal self-restraint and the foundation of any society, guarantees security and fairness and serves the public interest in personal relationships, in the sphere of morality, and in a political economy. Natural liberty and private property, the bases of a laissez-faire economy, make sense only in the framework of the restraints of justice.

That Smith did not talk about sympathy, benevolence, or the impartial spectator in the *WN* may have been a deliberate oversight. It is difficult, at best, to apply these notions to Smith's economic scheme. Self-interest is not antithetical to benevolence, but reciprocity, cooperation, and justice rather than benevolence are central to economic activities. The reason is that although economic activities in a political economy should ideally follow the precepts of fair play, when they do not they can be regulated universally and impartially. It is these qualities emanating from the social passions and interests rather than sympathy or benevolence that play a central role in exchanges and thus exchangeable value. The invisible hand is not patterned after the impartial spectator, and to attempt to draw such an analogy erroneously personifies the invisible hand. Here indeed, the *TMS* and the *WN* differ, not because they contradict each other, but because each is dealing with what Smith thought to be a discrete subject matter. Perhaps Smith thought that the notions of self-interest, the social passions and interests, and justice were general enough to enable him to link the two works. To hazard more would be philosophically questionable.

Some questions remain. First, though challenging many of the tenets of what I have characterized as a number of versions of a self-interest view of Smith, I have not yet considered whether he was a radical individualist, as some commentators claim. What is Smith's view of the relationship among individuals, the social order, and the institutional framework of a political economy? Second, even if various versions of the self-interest view are exaggerated readings of what Smith intended in the *WN*, what is the status of the economic model he presents? Is it a utopian ideal, or did

Smith think the model could be actually achieved? Is Smith optimistic about the future of laissez-faire capitalism, or does he see its pitfalls, which may portend its doom? Third, one must respond to Warren Gramm's conclusion that the *WN* is an eighteenth-century text written for the economy of that time. What it has to say is historically important and influential, but its conclusions are relevant only in the framework of an eighteenth-century precorporate entrepreneurial economy in the infancy of the Industrial Revolution. As such, Gramm concludes, the arguments in the *WN* do not and cannot apply to twentieth-century economic theory or practice. These all are topics for the next chapters.

# 4

# Individualism, the Social Order, and Institutions in Smith's Political Economy

> The really significant concept in the economic analysis [in the *Wealth of Nations*] is the concept of individuality, as inclusive of both self-interest and competition. Upon this conception the whole order rests. Adam Smith conceives of the economic order as purely a collection of competing individuals.[1]

This quotation from a well-known commentator on Adam Smith, Glenn Morrow, illustrates one version of the self-interest view of the *WN*. It attributes to Smith a form of economic egoism, or at least non-tuism, coupled with a radical individualism such that "the ideal system is one of fragmented economic power in which self-interest leads participants to cooperate with each other in transactions which yield mutual gain."[2]

In the last chapter I questioned the thesis that self-interest is the dominant motivating force in Smith's depiction of a political economy in the *WN*. In this chapter I shall focus on his alleged individualism. Morrow's reading of the *WN* forms part a view that is widely accepted, and more recently it has been qualified in the writings of Lionel Robbins and challenged by Nathan Rosenberg and Warren Samuels.[3] To quote Rosenberg, a "neglected theme running through virtually all of the *Wealth of Nations* is Smith's attempt to define, in very specific terms, the details of the institutional structure which will best harmonize the individual's pursuit of his selfish interests with the broader interests of society."[4] Or according to Samuels, in the *WN* "self-interest exists only within social control."[5]

I shall challenge Morrow's analysis, and at the same time I shall raise some questions about Rosenberg's and Samuels's more institutionalist reading of the text. Morrow may confuse Smith's alleged non-tuism with individualism, and Rosenberg and Samuels sometimes appear to overemphasize the role of self-interest in the *WN*. The existence of the social passions that are manifested in economic activities as cooperation and coordination precludes the conclusion that Smith's individualism is merely inclusive of self-interest and asocial competition. Robbins is correct when he writes that Smith presupposes a foundation of government, law, and social and religious institutions in which he locates his political economy. Smith does not imagine that a political economy could function without such support, and this crucial point is often neglected by commentators on the *WN*. But Rosenberg's use of the term *institution* is unclear, and sometimes he implies that Smith had in mind institutions that actually regulate self-interested economic activities rather than those that provide the framework of commutative laws of justice. I also question Samuels' agreement with Morrow that "Smith used two analytical procedures: the method of regarding society as a derivative of the individual *and* that of regarding the individual as a product of society"[6] and qualify Samuels' depiction of Smith's political economy as a dual and conflicting system of markets and power.

## Smith's Alleged Organicism in the *TMS*

In the *Theory of Moral Sentiments* (*TMS*) Smith writes: "Man, who can subsist only in society, was fitted by nature to that situation for which he was made. All the members of human society stand in need of each others assistance, and are likewise exposed to mutual injuries."[7] Glenn Morrow has argued that even though Smith retains the eighteenth-century tradition of individualism in the *WN*, in the *TMS* he departs from that tradition. "In contrast with the prevailing 'nominalism' of eighteenth-century social philosophy, the ethical theory of Adam Smith implies the existence of a social order possessing a natural and real unity."[8] As Morrow recognizes, a number of seventeenth-century philosophers, including Hobbes and Locke, held the view that "human individuals are the ultimate data of all human relations."[9] If individualism correctly characterizes the thinking of these philosophers – articulated in twentieth-century terms – these philosophers are "ontological individualists" because they hold that only individuals exist and have status as entities. In addition and more radically, these philosophers have sometimes been labeled "methodological individualists" who hold that social, economic, and political institutions are merely aggregate collections of individuals. None has an identity, character, or personality of its own, and none has any status apart from its members, constituents, or citizens. Each is derived from, and is reducible to, the sum of the individuals who make it up. Therefore, no institution acts as a collective or has responsibilities that supersede, or cannot be redistributed to, the individuals that constitute it. Whether or not Hobbes, Locke, and other seventeenth- and eighteenth-century philosophers are correctly called "methodological individualists" I shall leave for another occasion, but I maintain that this term does not describe Adam Smith's point of view in any of his texts.

Morrow finds that Smith perpetuates the philosophy of "abstract" or methodological individualism in the *WN*. Because the motive of individual self-interest dominates that work, according to Morrow, Smith focuses his economic analysis on individual exchanges. The free market is merely a collection of individual self-interested economic activities. "The state [is] merely a union of individuals for economic ends, . . . [and] every social phenomenon is explained in terms of the self-interest of individuals."[10] Justice, the only enforceable and codifiable virtue, replaces benevolence, allegedly the supreme virtue in the *TMS*, according to Morrow, that is necessary to control self-interests.

In contrast, Morrow continues, Smith develops a rich ethical theory in the *TMS* in which there is "a recognition of the social order both as the complete expression of the individual life, and as an organic unity maintaining itself independently of the will of individuals."[11] Unlike what Morrow calls the "abstract individualism" of the *WN*, Smith proposes in the *TMS* a more elaborate scheme in which the social order takes priority over the individual. Indeed, Morrow asserts, Smith's unique contribution to social science is his recognition that human beings are social beings who naturally are part of, and depend on, society rather than merely constituting and contributing to it.

Because, according to Smith, human beings are dependent on one another and have social as well as selfish passions, because sympathy depends on social relationships, and because moral judgments are derived from sympathy, the general rules, and the impartial spectator, Morrow concludes that morality has its roots in the social order.

> [Man's] very interests are those which have been instilled into him by his social environment, and hence in pursuing his own interests he is also pursuing those of his society. The rigid distinction between self and others is seen to break down . . . even the motive of self-interest is not individual but social in its origin. The *Moral Sentiments* shows the inner organic relation which exists between all the individuals of a society and the social unity.[12]

This interpretation of Smith, Morrow claims, accounts for the fact that in the *TMS* benevolence is the highest virtue, the virtue of social relationships. Thus in the *TMS* Smith breaks with the eighteenth-century tradition of individualism. It is the social order that dominates, and individuals are important insofar as they are part of that order. Again, using twentieth-century technical terms, according to Morrow, in the *TMS* Smith holds a theory of organicism or, in contemporary technical terms, "ontological collectivism."[13]

## The Role of Institutions in the *WN*

In contrast with Morrow, Lionel Robbins presents a "market plus framework" interpretation of Smith's political economy. Robbins is critical of the nineteenth-century "night watchman" theory of the *WN* in which Smith is depicted as a virtual political and economic anarchist, what in this century we might call a "libertarian." Smith's conception of liberty "did not exclude the conception of order,"[14] Robbins writes, and when Smith talks of "natural liberty" or "perfect liberty" he is referring to economic freedom operating within that order. It would be unthinkable, Robbins states, given Smith's eighteenth-century British background, that he would imagine a market system functioning without a constitution, a well-defined legal system and moral codes, and a national religion. One of the conditions for a perfect economic order is the institutionalization of laws of justice protecting individuals from harm and enforcing fair play. This framework is specifically articulated in the *Lectures on Jurisprudence* (*LJ*) and is presupposed in the political economy outlined in the *WN*. Smith's ideal of commerce is embedded in a larger religious, social, political, and institutional structure that establishes moral and legal rules and guidelines for economic behavior.

At the same time, according to Robbins, Smith, like all classical economists, was an individualist. Although we all are mutually dependent on one another,[15]

> the Classical Economists . . . were both individualists as regards ends and (with due reservations) individualists as regards means. For them, an organization of production, based, in the main, on private property and the market, was an essential complement to a system of freedom of choice as regards consumption and provision for the future. They believed that, within an appropriate framework of law, such an organization could be made to work harmoniously.[16]

Rosenberg and Samuels develop this analysis. "Adam Smith most distinctively stood for private enterprise, private property, self-interest, voluntary exchange, the limited state, and the market. . . . The market . . . is a mechanism for resolving basic economic problems and for producing order without elaborate central direction."[17] Nevertheless, Rosenberg sees two important theses in the *WN* that he finds often ignored by contemporary readers. First, although every person is motivated by his or her own self-interest in the market, this motivation has two sides. Each of us seeks to "better our condition," and Smith finds that each of us is the best judge of how to develop our economic interests, but it is also true that "it is in the interest of every man to live as much at his ease as he can."[18] This latter interest leads to loss of parsimony, slothfulness, indolence, and laziness, particularly in those who are economically successful, but none of us is exempt from these vices. Accordingly, evidence of imprudent self-interests leads Smith to inquire into what societal structures would best regulate these self-interests to enhance the social good. According to Rosenberg, Smith felt that competition was not enough to control economic activities and that we need to "harness man's selfish interests to the general welfare."[19]

At the same time, as Rosenberg correctly points out, Smith is critical of many of the institutional arrangements in eighteenth-century England. One of the purposes for Smith's writing the *WN* was to attack and present a remedy for the prevalent economic theory and practice of eighteenth-century Britain and the European continent, a theory and set of practices Smith terms *mercantilism*. In fact, mercantilism was a set of disorganized and often contradictory economic regulations and practices. However conceived, the English government perpetuated and imposed a number of restrictive regulations on economic activities designed to conserve the wealth of that nation. These regulations controlled agriculture and manufacturing, restricted exports and imports, and even monitored the movement and wages of labor. The result was that some forms of agriculture and industry were favored over others, thereby creating both monopolies and shortages; imports were virtually prohibited; and exports of certain goods were encouraged. Smith questions these practices because they restrict economic freedom, many favor one economic group over another, and they neither contribute to nor conserve wealth.

Extrapolating from Smith's critique of eighteenth-century governmental institutions, regulations, and laws governing economic activities, Rosenberg contends that "Smith was very much preoccupied with establishing the conditions under which this [free] market mechanism would operate most effectively . . . an exact, detailed specification of an optimal institutional structure."[20] Rosenberg cites Smith's statement that "publick services are never better performed than when their reward comes only in consequence of their being performed, and is proportioned to the diligence employed in performing them"[21] as the "guiding principle in the organization of public affairs."[22] For example, Smith points out, teachers and clergy should be paid for their services by their students or parishioners who will pressure these individuals out of indolence into good performance.

But what is worrisome is Rosenberg's use of the term *institution*, which is not quite the same as Robbins's use. Robbins describes an underlying system of religious, social, and judicial institutions as a background for Smith's ideal political economy, but Robbins does not focus on particular institutions such as the church or universities. Because of his criticism of mercantile regulations, Smith seems less concerned with specific organizations or mechanisms than with proposing a more spontaneous market-based arrangement for religious and educational instruction.

Samuels develops and extends Rosenberg's thesis. Citing Smith's arguments that a nation must provide education for everyone as one way to balance the degradation of the division of labor,[23] Samuels expands this analysis to argue that in general, according to Smith, "institutions . . . govern the answer to the question of whose liberty is to be achieved. . . . The individual is in one sense the prime element or unit in the economic system, but the individual exists and acts only within the evolving moral, legal, and institutional framework as a socialized individual."[24] Consistent with Robbins's analysis with which he agrees, Samuels finds Smith's underlying framework to include not merely constitutional and legal protections but also such nonlegal forces as religion, customs, morals, and education. Writing in 1966, Samuels finds that Smith assumed that these sorts of institutions would come to bear on an economy and economic activity in ways perhaps even more important than the law would.[25] In later articles, however, Samuels dwells on more formal institutional arrangements as crucial to governing economic activities.[26]

Within this societal and institutional framework Samuels sees Smith as holding two overlapping and sometimes conflicting models of the economy: the market economy and the "economy as a system of power."[27] According to Samuels's reading of Smith, although individuals ideally should have the natural liberty to exercise their own economic choices, this free-market model does not exhaust the description of economic interchanges. Economic freedom operates within the structure of rights and power. Competition includes a struggle for labor, land, and resources, and this struggle entails what Samuels calls "mutual coercion." Smith sometimes defines wealth as the power to command labor,[28] and Samuels sees this as economic power, which leads to conflicts between laborers and manufacturers.[29] These conflicts develop into forms of mutual coercion, because the laborer cannot live without work, and the manufacturer cannot succeed without labor. There is a similar struggle between manufacturers and landholders, both of whom compete with each other and struggle for power.[30] Samuels reads Smith as granting the government the task of encouraging a competitive free market, protecting property rights, and safeguarding the system of power. Laws of justice protect contracts and fair play and at the same time guard property rights, which, Smith admits, give power to the rich over the poor. Legislation that tries to balance these interests usually favors one over another and therefore is unacceptable to Smith. So, according to Samuels's reading of Smith, the competitive forces of market exchanges operating within, or in contraposition to, mutual coercion balance or are a countervailing force to the system of power. Superimposed on this dual picture of the economy is the control exercised by societal institutions that work to correct excesses in economic behavior, both competitive behavior and power struggles.[31]

In the *TMS* Smith argues that human beings are social beings and that morality exists only in the context of social relationships. Despite Morrow's arguments to the contrary, in Chapter 3 it was shown that this social context is understood in the *WN*, because economic activities involve competition and cooperation, and that these entail social interrelationships that depend on a well-ordered and stable social and institutional framework that secures and protects those activities. As Smith explains in the *Lectures on Jurisprudence*, private property develops in and is dependent on the framework of a civil society in which property rights are defined and protected. Stability, security, and law are necessary for any operative economy. Smith's development of a political economy is the development of such a framework in which a viable economic system can become established and mature. Furthermore, implicit in Smith's analysis are his roots in eighteenth-century British Protestantism that provide the conditions for minimally acceptable moral behavior, the behavior expected of Protestant British gentlemen. Robbins's description of the function of institutions in Smith's political economy, then, seems accurate.

One must be cautious, however, in extending an analysis of the institutional and social framework of Smith's political economy beyond Robbins's outline. In the *WN* Smith recognizes the importance of institutionalized laws of justice: "Every man, as long as he does not violate the laws of justice, is left perfectly free to pursue his own interest his own way, and to bring both his industry and capital into competition with those of any other man, or order of men."[32] As he argues in the *TMS* and the *LJ*, such laws can be spelled out,[33] and he finds such laws necessary to protect property and property rights and to control some economic self-interests and cooperative ventures that are not self-restrained by a sense of prudence, public interest, or fair play. As he carefully specifies in the *LJ*, however, this system of justice is a system of commutative justice composed of laws to protect what he calls perfect rights. Forms of distribution or reallocation that favor one party over another do not apply equally to everyone, nor are they required by natural law, and so such distributions are to be left to charity and the benevolence of individuals.[34] The ideal political economy, then, protects economic liberty, provides security for the society, and institutes laws of justice to ensure fair play, but it should not engage in economic manipulations designed to distribute or redistribute economic goods, services, property, labor, or wealth.

Rosenberg implies that Smith seeks to provide a positive secondary set of institutions that would support an efficient market, govern human selfish behavior, and promote the public good. But it is not clear what sorts of institutions Rosenberg has in mind. In addition to the framework that Robbins attributes to Smith, the market itself, market exchanges, and a "laissez-faire" price system constitute some of those institutions. Rosenberg also refers to "appropriate institutions" surrounding these economic ones that "harness man's selfish interests to the general welfare."[35] But there is little textual evidence to confirm what, for Smith, might constitute an institutional framework beyond that specified in a system of constitutional commutative justice that would regulate what Rosenberg finds to be overindulgences in economic self-interest. The passages that Rosenberg cites as evidence of Smith's aim cannot serve as evidence of Smith's interest in a more specific positive institutional framework for governing economic activities, because what he is advocating there is the abolition of an institutional or governmental system for the remuneration of public servants and public services and replacing that with incremental pay and piecemeal reimbursement, depending on the quality and quantity of service rendered or the use of a public work.[36] This is hardly what one would call an "institutional arrangement" but, rather, an alternative to such arrangements. So although it may be true that an institutional analysis of the *WN* has largely been neglected, it is more difficult to find textual evidence for a more detailed specification of the sorts of secondary institutions that Smith might have in mind. Perhaps Rosenberg is using the term *institution* in a weaker sense to refer to the social, political, and religious framework in which economic activities are viable. But it is more difficult to find evidence of Smith's search for an "optimal institutional structure" in the *WN* that will govern self-interest and promote commerce beyond the framework that Robbins outlines.

Rosenberg's reading of self-interest as the guiding principle of the *WN* may lead him to interpret Smith as being in favor of specific institutional structures beyond those of the market, moral and religious taboos, and laws of justice to guide our self-interested and indulgent activities. But self-interest is not the only basis for economic exchanges, and prudence and cooperation play equal roles, as I argued in the preceding chapter. The conclusion that competition within the framework of fair play is sufficient to control imprudent and indulgent behavior is a reasonable interpretation of the *WN*. Smith's persistent critique of institutional regulations because of their favoritism and inefficiency, a critique on which Rosenberg elaborates, and the fact that Smith explains little about the positive institutional structures that both exclude special privilege and guarantee competition lead me to conclude that Smith favors a framework of law and rules of justice that interferes little with the activities of commerce.

## Methodological Collectivism in the *WN*

Samuels states that "Smith . . . blended methodological individualist and methodological collectivist levels of analysis. . . . Smith used two analytical procedures: the method of regarding society as a derivative of the individual *and* that of regarding the individual as a product of society."[37] To understand what is at stake in this statement, let us first briefly explore what is meant by "methodological collectivism," "individualism," and "organism," terms that

Morrow uses to describe Smith's views. I shall then investigate whether and in what sense Smith is a collectivist in the *WN*. Next, I shall reexamine the *TMS* to see whether Smith truly argues against individualism in favor of a form of social organic unity or holism, as Morrow contends.

An individualist maintains that "individuals are the ultimate data of all human relations." Given that general definition, individualism can be interpreted in at least two ways, as I pointed out earlier. First, one can hold the view that only individuals exist, that is, that there are no "super entities" such as institutions, societies, legal systems, or nations that exist apart from the individuals who make them up or to whom the laws or mores apply, or "ontological individualism." Second, one can hold a stricter form of individualism, or "methodological individualism." Not only do only individuals exist, but also all actions, facts, explanations, laws, and other data of the social sciences can be reduced to facts and explanations concerning individuals and individual behavior.[38]

Some social scientists and philosophers believe that although the ultimate constituents of the world are individuals, individuals create institutions, legal systems, and social structures that take on a character or personality of their own. These "societal facts" or "social objects" are created from, exist only because of, and can be changed or annihilated only by individuals. Nevertheless, as "collective conglomerates" created from the input of various individuals and groups of individuals, "societal facts . . . which are used to refer to the forms of organization of a society cannot be reduced remainder to concepts which only refer to the thoughts and actions of specific individuals."[39] Groups, institutions, nations, and social or legal systems act as individual entities, and we treat them or respond to them in that way. They engage in collective action that cannot be descriptively reduced to merely the sum of individual actions of their participants, and laws may be drawn up to govern their behavior. They are theoretical entities that have explanatory force, although ontologically they have no status. Or sometimes they are thought of as conventions that expedite and explain social action and coordination.[40] This set of views is often called "methodological collectivism" or "moderate collectivism." A fourth view, "ontological collectivism" or "ontological holism," sees collective entities as having an independent and autonomous existence apart from, or despite, individuals. Some ontological holists then argue that individuals are a part or a product of, or are derived from, such societal entities.[41]

Samuels is suggesting that Smith sometimes appears to be a methodological individualist, thus an ontological individualist and sometimes a methodological collectivist, particularly when Smith talks about civil government and laws of justice. Samuels emphasizes that "institutions, etc. are the product of the aggregation of individual human action (there are no other actors but humans)."[42] Samuels also claims, however, that he agrees with Morrow that "Smith used two analytical procedures: the method of regarding society as a derivative of the individual *and* that of regarding the individual as a product of society." The latter is not methodological collectivism, because the methodological collectivist believes that although the notion of society may have explanatory force, society and societal institutions are products of individuals. Individual persons operate within, are influenced by, and react to societal facts, but individuals are never a product of those facts. What Samuels appears to attribute to Smith in this passage is a form of ontological collectivism similar to Morrow's attribution of organicism to Smith in the *TMS*, in which individuals are derived from and a product of society. This appears to be inconsistent with the rest of Samuels's interpretation of the *WN* and with Smith himself, as I shall now argue.

## Self-Interest, Institutions, and Collectivism in the *WN*

The role of institutions in the *WN* is both more simple and more complex than Morrow, Rosenberg, or Samuels describes them. I maintain that the exaggerated claim that self-interest is the sole motivating element in economic exchanges forces Morrow to ascribe to Smith a radical individualism on the one hand, but this view encourages Rosenberg to read into the *WN* an overemphasis on Smith's preoccupation to establish an institutional framework to control self-interests. This emphasis on self-interest leads Samuels to the interesting conclusion that Smith's depiction of a political economy is, in part, an "economy as a system of power." I shall take up this last issue at the end of this chapter.

In the *WN*, self-interest is not the sole or even the principal motivating force in economic exchanges. There are two reasons that this is so, as I argued in Chapter 3. First, it is not merely self-interest to better our condition that motivates us but also a desire for respect and admiration. Second, economic interchanges depend on cooperation and reciprocal agreements that are triggered by the social passions, as it is the social passions that motivate and prompt our interest in, although not necessarily our love or approval of, others.

If self-interest is not the sole motivating force in economic exchanges in the *WN*, then it does not follow that Smith is necessarily an individualist in that text. But one must be careful to qualify his alleged "institutionalism" or collectivism. Because of the role of the social passions and interests in economic affairs, the importance of a complex institutional framework to harness self-interest may not be as critical as Rosenberg seems to believe. In the *WN*, Smith recognizes the genuine existence of institutions, but he did not imagine that individuals and individual behavior could be merely part of an institution. For him, all actions can ultimately be traced to the actions of individuals, and the activities of institutions, states, or nations are ultimately a collection of the actions of individuals who make up those collections.

A number of passages in the *WN* support this conclusion. Although the character of laws and institutions constitute some of the conditions under which individuals function and function well—as amply illustrated in Smith's description in Book IV of mercantilism and its effects on economic activity—these laws and institutions evolve in the first instance incrementally by individuals and can be changed accordingly. This is evident from Smith's analysis of the development of nations from a rudimentary hunter stage to advanced commerce. Moreover, how individuals and, in particular, leaders interpret and carry out institutional guidelines such as laws affects in large measure the resulting societal consequences. Smith's criticisms of the indolence of the rich, the laziness of schoolmasters and professors, and the slothfulness of judges and the clergy demonstrate this point.[43]

Smith's individualism carries over to his analysis of a political economy and a commonwealth. Although he stresses the importance of laws of justice, these laws are negative precepts protecting people from harms and guarding fair play. Because Smith is careful to restrict the duties of government to protecting citizens from invaders, enforcing laws of justice and contracts, building public works in everyone's interests, and education, this is hardly the kind of collectivist institutional framework that could regulate the market.

It is plausible, however, to argue that Smith is a methodological collectivist in the sense in which I have defined it and in the way in which Samuels points out, except for that one unfortunate passage in which he concurs with Morrow. Smith implies that institutions do develop their own character. At the least, he recognizes that we often treat institutions as if they had their own explanatory power. This is evident when he talks about the role of laws and institutions in China and other foreign countries.[44] It is also clear in Smith's genuine fear of institutions, as shown in his critique of the system of mercantilism, of monopolies, and of political or economic institutions that favor some individuals over others.[45] Smith questions the existence of "joint-stock companies" (corporations), except in exceptional circumstances, because the institutionalization of management power separated from ownership creates institutional management power cut loose from responsibility.[46] Smith's fear is that such institutions might become personified, so that one would regard them as real entities and hence treat them as incapable of being dismantled.[47] As I argued in Chapter 3, even in his analysis of the invisible hand, Smith is careful not to reify that phenomenon. Although he undoubtedly assumes a background of social and political institutions that form the support of his political economy, his criticisms of mercantilism and of other regulations imply that he is wary of institutional interference. Often he opts for the evils of unfair competition instead of the remedy of law or regulation,[48] because he recognizes that institutions do take on "personalities" of their own that sometimes translate into power and the misuse of power. It is not erroneous, then, to call Smith a moderate methodological collectivist, but he is clearly not an ontological collectivist.

If it is the case that the ideal framework for a political economy guarantees security, protects economic liberty, and adjudicates fair contracts and other practices but interferes little with the workings of an economy, can one explain how an economy "will best harmonize the individual's pursuit of his selfish interests with the broader interests of society?" For Smith, unlike his interpreters, this explanation is not so difficult as it would seem. There are a number of reasons for this. First, as I have argued, economic actors are not motivated merely by self-interest, and a well-run competitive economy depends equally on cooperation. Because the individual pursuit of economic betterment is not identified with the unilateral pursuit of self-interests, Smith does not have to postulate a secondary institutional framework that controls these interests. Second, for Smith, problems that prohibit competition include, but are not exhaustively defined by, greed, sloth, laziness, or other vices of self-interest. He appears to assume that most of us will act prudently a good deal of the time, thereby emulating the virtues of Protestant British gentlemen.[49] Cooperative ventures, too, are often collusive or do not take in account the public interest. The real difficulty, Smith finds, is that in general, human beings are shortsighted as well as unconcerned, and it is the shortsightedness of individuals and groups rather than their asocial self-interest that is often not in the public interest. Laws of justice, then, Smith implies, provide the framework to safeguard fair play, and under the conditions of fair play, the pursuit of competitive and cooperative interests ordinarily, although often unintentionally, enhances public welfare. Except in extraordinary circumstances, such as in the banking industry in which piecemeal regulation may be necessary to counteract monetary dangers to society,[50] Smith appears to think that a well-formulated system of commutative justice coupled with extralegal moral forces of prudence, cooperation, and a system of public education are enough to regulate private interests while allowing economic liberty to flourish in the pursuit of public welfare as well as private well-being.[51]

## Collectivism in the *TMS*

If Smith is a moderate collectivist in the *WN*, is he a more radical collectivist in the *TMS*, as Glenn Morrow argues? Morrow rightly demonstrates that in the *TMS*, sympathy, moral judgments, and even Smith's notion of conscience depend on social relationships. As Smith states:

> Were it possible that a human creature could grow up to manhood in some solitary place, without any communication with his own species, he could no more think of his own character, of the propriety or demerit of his own sentiments and conduct . . .

> than of the beauty or deformity of his own face. . . . Bring him into society, and he is immediately provided with the mirror which he wanted before.[52]

Yet it is an error, I think, to conclude with Morrow that in the *TMS* the social or institutional order has an organic unity that preempts individual identity. That Smith is primarily an individualist is clear in all his writings. Although he says that "man . . . can subsist only in society,"[53] he develops no concept of a society, collectivity, or collective action apart from the sum of individual actions. The selfish, unsocial, and social passions, important concepts in both the *TMS* and the *WN*, focus on the ultimate importance of individuals rather than on institutions, the commonwealth, or the state.

To review, in the *TMS*, Smith is concerned about the individual, his or her passions and interests, conscience, and virtues. Even justice in the *TMS* is primarily an internal mechanism of control. While recognizing that in the trivial sense all our passions are self-interested because they originate in ourselves, in the *TMS* Smith argues that human beings "by nature" are motivated by three sorts of passions. He gives credit to the social passions as the source of interest in others and thus as a source of sympathy and benevolence. The fact that we have social passions and display sympathy implies that our passions and interests are not asocial, or purely selfish. The social passions, however, are individual passions. They are genuinely other-directed, and we care about and need others. But the existence of the social passions and our need for others do not necessarily imply a superimposed social order. A "social order" is merely a collection of social relationships among individuals. These social relationships are natural, just as self-interest is natural. What is unique about Smith's moral psychology is that he makes a good case for the claim that individuals are "by nature" both self- and socially interested without committing himself to either a form of egoism or a kind of organicism or holism.

Smith places social relationships on a par with other individual relationships, but a notion of the organic unity of a social order reads too much in his more modest claim that human beings are naturally social. Just because there are societal norms and institutions, this does not militate against individualism, as Robbins points out. Smith makes clear in the *TMS* that such norms and institutions are a product of activities of sympathy and the impartial spectator and also function because of, and are reducible to, individual activities. He explains:

> The concern which we take in the fortune and happiness of individuals does not, in common cases, arise from that which we take in the fortune and happiness of society. . . . [O]ur regard for the multitude is compounded and made up of the particular regards which we feel for the different individuals of which it is composed.[54]

In writing about the relationship between political leaders and the state, Smith remarks, "In the great chess-board of human society, every single piece has a principle of motion of its own, altogether different from that which the legislature might chuse to impress upon it."[55] In the *TMS*, then, Smith is an ontological individualist.

Although Smith would not make that distinction, is he a methodological collectivist in the *TMS*; that is, do societal laws and institutions have their own character and explanatory power or, at least, a conventional role? In the *TMS* he talks little about institutions, and society appears to be an aggregate composition of individuals and groups: "Upon the manner in which any state is divided into different orders and societies which compose it, and upon the particular distribution which has been made of their respective powers, privileges, and immunities, depends, what is called, the constitution of that particular state."[56] On the other hand, when Smith talks about nature or the natural order he writes as if nature, as a unifying or universally present phenomenon, has explanatory force. Indeed, sometimes Smith talks as if nature has an organic unity underlying individuals and individual order.[57] And it is unclear whether nature is the foundation for human society or the ideal to which we innately strive. Nevertheless, even in the natural order, individuals are primary: "Every man is, no doubt, by nature, first and principally recommended to his own care."[58] Therefore, there is little evidence of methodological collectivism in the *TMS*.

## Markets, Coercion, and Power in the *WN*

Finally, let us return to Samuels's reading of the *WN* as depicting a system of free markets, restrained by both a countervailing system of mutual coercion and superimposed institutional societal control.

I have two questions about this reading of the *WN*. First, Samuels does not develop the important fact that for Smith the market is an outcome of individual activities, and thus individual economic actors play crucial roles in the kind of market that develops and in the way that it operates.[59] So individualism is important to a description of the market.

As illustrated in the passages that Samuels cites, I too see Smith as struggling with the notion of power. The bases for competition—private property and natural liberty—create wealth, which in turn creates power. Merchants and manufacturers do exploit labor; landholders do struggle with manufacturers for power. But I am not sure that "Smith understood the economy as a system of mutual coercion,"[60] as Samuels claims. I see Smith as focusing on three contributory forces in a free economy. First, we as individuals, enjoying natural liberty to pursue our economic ends, are motivated by our selfish and social passions and thus seek to improve our condition and to cooperate with and receive the approval of others. At the same time each of us overindulges, colludes with neighbors or business acquaintances, takes advantage of laborers or renters, and often feels envy and other unsocial passions toward others. The confluence of these passions accounts for the fact that we are both competitive and cooperative and also that we are not always impartial. Private, but not necessarily selfish, interests often take precedence over public interests, non-tuism over disinterested concern.[61] Thus, according to Smith, because of our shortsightedness, a system of justice is necessary to protect a sense of fair play.[62] Second, for the same reasons, a complex regulatory system is also not in the public interest, because the same sort of partiality is likely to be exhibited by those working in the public sphere. Third, a system of natural liberty creates property and inequalities of wealth and thus inequalities in economic power.

Smith recognizes that "wherever there is great prosperity there is great inequality. . . . The acquisition of valuable and extensive property, therefore, necessarily requires the establishment of civil government."[63] Laws of justice are needed "for the defence of the rich against the poor, or of those who have some property against those who have none at all."[64] But Smith does not work out how the propertied and the propertyless balance or coerce each other. Rather, he finds that in a mercantile economy, attempts to orchestrate economic affairs give advantages and thus power to some while setting up roadblocks for others. The result is a creation of monopolies for which there is no countervailing control or power, protections of employers from employees without any reciprocal protection of workers, trade barriers that hamper rather than enhance the economy, and favoritism for some multinational enterprises at the expense of domestic businesses. These economic power bases, Smith argues, are best dissipated not by institutional restrictions but by a competitive market economy backed by a well-formed system of justice. In his thinking, almost no law that specifies the kind or quantity of economic exchange is equally fair to everyone, thus such laws should be avoided.[65] Moreover, and again perhaps naively, Smith finds that competition, even among unequal parties, seldom gives monopolistic economic power to one individual or any one group of individuals for very long. A commercial economy, however, makes everyone relatively powerless, rather than creating a system of countervailing power or mutual coercion by dispersing economic advantages.[66] Such an economy replaces power not by a system of mutual coercion but by a system of competition in a framework of laws of justice. Therefore it is the institutionalization of laws of justice, not a system of mutual coercion, that best preserves individualism as well as economic liberty in an ideal political economy.

# 5

# Labor, the Division of Labor, and the Labor Theory of Value

> The annual labour of every nation is the fund which originally supplies it with all the necessaries and conveniences of life.[1]

One of the most important notions in the *WN* is Smith's concept of the division of labor. Concerning its centrality in that text, Joseph Schumpeter writes: "Though, as we know, there is nothing original about it . . . nobody, either before or after A. Smith ever thought of putting such a burden upon division of labour. With A. Smith it is practically the only factor in economic progress."[2] The concept of the division of labor is an ancient one, and its role in economic growth was already a matter of discussion in Smith's time.[3] The dividing of tasks exists in the most primitive societies, and earlier thinkers such as Locke noticed that it is human labor that makes a difference in economic value. Smith, however, makes labor, the labor theory of value, and the division and specialization of labor the cornerstone of his economic theory. He states that "the greatest improvement in the productive powers of labour, and the greatest part of the skill, dexterity, and judgment with which it is any where directed, or applied, seem to have been the effects of the division of labour."[4] Labor, he argues, is an important source of wealth, because only labor can increase its productivity and thus create wealth. Because labor has invariable value to the laborer, labor is the best measure of the real value of commodities. In a commercial society in which there is a demand for goods, the organization of productivity—that is, the division, specialization, and mechanization of the work force, along with capital or stock and land or rent—are the driving forces of such an economy. Thus the division of labor is the source and measure of new wealth, economic growth, and therefore economic well-being.

The importance of the division of labor in Smith's economic theory cannot be exaggerated. However, in this chapter I shall argue that Smith's unique contribution to the analysis of labor is his thesis that economic liberty develops out of the distinction among laborers, laboring, and their productivity, that the productivity of laboring, but not the laborers, is treated as an exchangeable value and therefore as a commodity. This distinction arises in situations in which at least some members of a community are relatively affluent and when their material demands cannot be met under present market conditions, demands that can be supplied only with increased production dependent on specialization of labor. The specialization of labor in turn, Smith explains, provides an important condition for economic growth and general well-being. More importantly, the separation of laborers from their productivity is the basis for worker freedom. But it is only when labor productivity is valued as a commodity that laborers are self-consciously differentiated from their productivity, and thus it is only then that laborers attain economic liberty.

Despite its central role in his political economy, however, there appear to be some paradoxes in Smith's theory of labor. First, he seems to have difficulty reconciling the goal of worker independence with the necessity of property and capital in a commercial economy. Smith is an advocate of economic liberty for laborers; he is highly critical of apprenticeships; and he favors the freedom of movement and occupational choices for all workers. He also believes that high wages both are fairer to the workers and provide incentives for better productivity. But he notices that manufacturers collude with one another to depress wages while wishing to increase productivity and that in general, the laborers are dependent on their employers for their wages and benefits. Smith recognizes that in a commercial economy laborers can never be paid fully for their contributions, and he worries about the effect of high wages on profits. Can laborers ever be truly free in face of the power and control of employment, property, and capital in the hands of others in a system based on property inequalities in which profitability, not high wages or worker freedom, is the primary concern?

Second, Smith acknowledges that the specialization of labor may harm laborers. In Book I of the *WN* he elaborates on the positive benefits of the division of labor, but in Book V he worries that overspecialization "corrupts the courage of his [the laborer's] mind . . . and even the activity of his body."[5] In Book I Smith finds that the division of labor leads to creativity and inventiveness, at least on the microlevel of specialization, but in Book V he finds the laborer becoming "stupid and ignorant." The question of the laborer's mental deterioration in an industrial setting is connected with another issue. Smith sees the distinction between the laborers and their productivity as necessary for both economic progress and the laborers' liberty. But does the concept of labor productivity as a commodity, plus the corruption of the laborers in an industrialized economy, lead to alienation? Does it also result in treating laborers as commodities?

Because of these apparent paradoxes in Smith's theory of labor, he is sometimes described as having anticipated Marx's analysis of labor and Marx's critique of the alienation of labor in a capitalist industrial economy.[6] I shall argue that Smith's analysis of labor anticipates standard Marxist treatments of the subject, by presenting a resolution to the problem of alienation in a commercial and industrial economy. Although Marx would not accept Smith's solution, it is likely that Smith would see his own analysis as an anticipatory response to Marx.

## The Commodification and Freedom of Labor

According to the historian Leonard Krieger,

> Adam Smith's labor theory of value was designed, not to favor industrial producers, but to indicate the productivity of every economically active individual, whatever his activity. He meant to emphasize the equality of all classes in the exchange process . . . and he meant to emphasize the economic creativity of the individual human being.[7]

A key to understanding Smith's theory of labor is his analysis of productivity as a commodity and its relationship to liberty and the division of labor. About the relationship of labor to the laborer Smith comments:

> The property which every man has in his own labour, as it is the original foundation of all other property, so it is the most sacred and inviolable. The patrimony of a poor man lies in the strength and dexterity of his hands; and to hinder him from employing this strength and dexterity in what manner he thinks proper without injury to his neighbour, is a plain violation of this most sacred property. It is a manifest encroachment upon the just liberty both of the workman, and of those who might be disposed to employ him. As it hinders the one from working at what he thinks proper, so it hinders the others from employing whom they think proper.[8]

In arguing that labor is a property, a "sacred property," Smith implies that the labor of the laborers, their "strength and dexterity," is inviolably theirs. His term the "original foundation of all other property" appears to refer to Locke's thesis that in a state of nature the enhancement of labor increases the value of unowned property, and because it is one's own labor, its "fruits," the enhancement of what one produces, including its products, are also one's own. However, as you may recall from Chapter 2, in the *Lectures on Jurisprudence* (*LJ*) Smith states that property is both an adventitious and a perfect right; that is, it is an acquired right whose violation requires correction. How, then, is labor a sacred and inviolable property and also an adventitious right?

In the passage just quoted, the expression "the property which every man has in his own labour" appears to refer to a natural right and also to label property as an adventitious right. But this is not an accurate reading of Smith's intentions. The confusion occurs because he, like Locke, does not clearly distinguish among the laborers, their laboring ("strength and dexterity"), and the productivity or "fruits" of their labor. As both the activity of laboring ("strength and dexterity") and its "fruits" or productivity, labor belongs to the laborer. One's labor, whether it is laboring or its productivity, is the "original foundation of all other property" because in the early stages of economic development, private property rights—for example, the rights of hunters to their kill, shepherds to their herds—are generally identified with rights of acquisition and accumulation primarily achieved by individual labor. But there is a crucial distinction between laboring and its productivity. One's "strength and dexterity" are "sacred" because they are naturally, inviolably, one's own. Therefore, "the property . . . of [one's] own labour . . . is the most sacred and inviolable" because laboring is part of the laborer, a natural and perfect right. Productivity, however, is the quantitative and qualitative input of a laborer's "strength and dexterity" that produces goods and services. Thus productivity is the capability of strength and dexterity and in this sense is distinct from laboring. Tully maintains that these distinctions are made by Locke in the *Second Treatise*.[9] Even if this is true, Smith extends this analysis. It is, at best, not clear that productivity is an alienable property, according to Locke. But Smith argues in another passage in the *WN* that productivity, even though it is the property of the laborers, is distinct from the laborers and their laboring; that is, it is an alienable property that can be bartered without harm to the natural rights of the laborers.[10]

One's productivity, as alienable property, is also "sacred and inviolable" in the sense that like all private property, the laborers have a perfect right to it. But productivity is also an adventitious right, because if productivity is exchangeable—according to Smith's technical definition of property in the *LJ*—the productivity of labor is an acquired right. Like other properties, Smith believes, productivity is not recognized universally as a distinct and exchangeable commodity; it depends on civil government to be protected; and its exchangeable value varies under different employment and market conditions. Because the productivity of labor is a marketable property and because labor varies in the quantity and quality of its productivity, productivity has value as a commodity. Because that property is one's own, to which one has a perfect right, and because productivity is exchangeable, one should be free to exchange this commodity, and others should be free to employ it. Thus one can sell one's labor productivity (but not one's strength and dexterity) without thereby selling oneself into serfdom. If one is not paid for one's productivity, one's property rights will be violated. Worse, because one's productivity is an outcome of one's own labor, if it is not recognized as an exchangeable commodity, one thereby will be treated as a slave.

The separation of productive labor from laboring and the laborers triggers both economic liberty and the valuing of labor. In any economy the realization that labor has value becomes apparent when productivity and efficiency as well as skill become either necessary or desirable. In a mature society in which the demand for goods exceeds what that society can supply by means of its traditional methods, such market demands initiate the separation of the laborers or tenants from their work or productivity. The fact that property is an acquired right explains this phenomenon. As the productivity of labor becomes valued as a commodity, it becomes a property with value to the laborers as well as to the employers. When labor becomes analyzed according to what it does or does not produce, labor's productivity becomes separate from the laborers, for it is the value of what is produced, both in quantity and quality, that is at issue, not the laborers themselves or their labor as such. Labor then has value as a commodity in terms of its productivity rather than merely as labor per se. The transformation of labor productivity into a commodity is progress, according to Smith, because the laborers, now being distinct from their productivity, are free to barter this commodity and, at least in theory, to determine how much they will command for it.

The valuing of productivity is an essential step in economic freedom and development. But, Smith explains, it derives from the existence of unmet market demands, followed by the division and specialization of labor, in which labor productivity can be increased as efficiently as possible. The specialization of labor enhances productivity and thus its value. The result of market demands and the specialization of labor is the concrete liberty of the laborers who are now independent, in control of their productivity, and its beneficiary. But without specialization, the valuing of labor and the freedom of the laborers assume a status quo; that is, the freedom becomes abstract, because the laborers are in a weak bartering position and so can neither exercise their freedom nor improve their economic position.

## The Development of the Specialization and Freedom of Labor

The foregoing is an abstract analysis of labor commodification and the independence of the laborer extrapolated from Books I and III of the *WN*. Smith's analysis, however, is more concrete. At issue is whether and how the commodification of labor productivity and the concomitant independence of the laborer are or can be achieved, and Smith endeavors to show how these take place: Without an unsatisfied market demand for goods, capital for industrialization, and the division and specialization of labor, labor freedom cannot be realized.

As we showed in Chapter 3, Smith finds that the division of labor is a "natural consequence" of our affection for, and need of, the assistance of others. The "propensity to truck, barter and exchange" arises from the social passions and interests and is in our self-interest as well, because by trading our surpluses or the produce of our labor, we are thereby able to increase our own advantages while cooperating with others. Thus the division of labor is natural; it derives from the social passions; and it is beneficial to ourselves as well as to society. A division of labor is found even in primitive societies. Hunters, for example, divide up the labor of slaughtering their kill. But in such societies tasks are divided for convenience. Labor is not valued or measured for its increased productivity, and any differences in contributions or surpluses of labor productivity are not thought of as such. In a primitive society in which the notion of private property is not yet fully or self-consciously developed, "the whole produce of labour belongs to the labourer."[11] In such a society, too, the laborers, their labor, and their productivity are not differentiated. But in such societies, labor and the productivity of labor have no value except to the laborers. Smith argues further than unless primitive societies develop an idea of private property beyond "fruits," such societies will not progress beyond the hunter or shepherd stage. Although the complexification of property is not an inevitable development, Smith contends that the development of property in the form of land and possessions is necessary for the expansion of herds and agriculture and, eventually, for commerce.[12] To review what we discussed in Chapter 2, the notions of property and property rights develop in the first instance from labor, because it is the expansion of herds and land development through acquisition and labor enhancement that create the notion of private property and, with it, inequalities. Despite this early origin of private property, however, there is not a concomitant development of the notion of labor productivity as alienable property until the age of commerce. Although Smith does not explain explicitly why this is the case, in the first three stages of his four stages of economic development, the primary focus is on agriculture, in which there is an abundance of laborers and little demand for goods that cannot be satisfied. It is only when the demand for labor exceeds the supply, Smith implies, that the commodification of labor productivity is realized.

In the stages preceding the age of commerce, the laborers, for example, the serfs, tenants or slaves, their labor, and what they produced often were valued as a single unit. When land was sold or traded, the serfs or tenants went with the land. It was the existence of a worker who worked that was of value, and as long as that person produced minimally, his work was not judged separately from his being a worker. By being identified with the laborer, labor productivity was valued as labor of the laborer, and distinctions between the quantity and quality of work were not always noticed and seldom rewarded.[13]

The development of private property beyond the fruits of one's labors in the precommercial stages does not challenge the identification of the laborers, their labor, and their productivity, and this identification precludes the recognition of productivity as a property of fluctuating exchange value. Smith is highly critical of the postproperty-precommercial conception of labor, because the laborers are identified with their productivity; labor is not recognized as a creator of economic value; and thus the laborers are not free. The preproperty inseparability of the laborers and their productivity is perpetuated in the third stage with a continued identification of the two in the mind of the employer, landlord, or sovereign, so that personal contribution and labor contribution are considered the same. The result produces three negative consequences: First, the laborers are treated as a commodity and thus demands are made on them as persons that exceed their work contribution, as demonstrated in the excessive demands made by the feudal lords on their tenants. Second, from the employer's point of view—as Smith notices in his discussion of slavery—the tenants, slaves, or serfs do only the minimum required, because they will receive the same, that is, subsistence, whether or not they are productive.[14] Third, the laborers, tenants, or serfs themselves have no conception of what it would mean to be at liberty to choose or change their occupation or barter their productivity.

Smith traces to the end of the feudal period the conscious development of labor as a value when landlords needed higher productivity from their tenants in order to support their increasing desires for luxuries. That is, the demand shifted from a demand for labor to a demand for productive labor, because in order to purchase new goods the landlords needed more productivity than traditionally had been available from their land. Tenants who worked efficiently became more desirable than others, and these desirable tenants were able to bargain for their labor. This bargaining power in turn gave them independence from a particular landlord as their protector and provider, independence from what was often arbitrary authority. Productivity thus became a commodity to be bartered for, exchanged, and traded, and labor and the productivity of labor, as distinct from the laborer or tenant, began to be understood as creating value. At the same time, the development of constitutional law, which Smith also traces through four stages, required that tenants be given leases, usually lifetime leases, for their right to work on the land. Under this arrangement, even tenants who were forced to pay full value to their landlord for the produce of their land were "freemen," according to Smith, because having leases for which they paid fully, they owed the landlord nothing, nor could the landlord demand more. The independence of labor from the arbitrary authority and protection of a single landlord gave the tenant a sense of what economic freedom might mean, as he was separated both from his landlord and from what he produced. The demand for luxury goods also put pressure on manufacturers and merchants to produce rather than import such goods, thus creating an expanding market for productive labor. Smith implies, then, that an unsatisfied demand for goods pressures labor for greater productivity, thereby giving productivity some value as a commodity. The laborers or tenants become valued or devalued not simply because they are laborers or tenants but for their labor contribution. This commodification of productivity separates laborers from their productivity and frees them from serfdom, thus securing their economic freedom. Moreover, an unsatisfied demand that creates a market for good labor is an important impetus for the development of commerce and a market system in which most goods and labor are freely traded.[15]

The valuing of productivity as a commodity and the development of markets initiate the age of commerce, Smith's fourth stage in economic development, and lead to the specialization of labor.

Although throughout the *WN* Smith uses the term *division of labor* to refer both to the division and the specialization of labor, he recognizes that the division and even the specialization of labor occur even in primitive economies. The self-conscious specialization of labor linked to its mechanization occurs only with industrialization, however. The latter, crucial to manufacturing, arises only when demands for consumer goods create a need to increase the productivity of labor output beyond what an individual or small group of individuals can ordinarily produce by hand. In addition to increasing productivity, saving time, and reducing labor costs, the specialization of labor allows the laborers to concentrate on one task or a series of tasks, thus eliminating the necessity for diverse skills of dexterity. By focusing on one aspect of the manufacturing process, the laborers can increase their skills in that task while in fact doing the job of many workers. Smith even argues in Book I that the division of labor and the narrowing of tasks give rise to creativeness and inventiveness by laborers who are able to envision improvements in their work.[16]

Because the division of labor can increase productivity and reduce the costs of goods, it is the source of economic growth and economic well-being. It allows the production of goods cheap enough to be purchased by laborers, and higher productivity, in turn, permits laborers to demand higher wages, which gives them a greater purchasing power with which to buy manufactured goods. That market demand, in turn, creates more jobs, jobs supplied by new manufacturing facilities made possible by the capital generated from previous industrial successes. As the demand for goods increases, so too does the demand for labor, and thus wages and the standard of living for that society go up.

Smith maintains that this process of economic growth cannot proceed initially without capital to purchase the necessary raw materials and machinery. Although he writes that "labor commands capital," he recognizes that capital is essential to begin the industrialization process. So the development of private property, commencing in the age of shepherds with the accumulation of herds, is necessary for capital accumulation and investment in the age of commerce. In advanced societies in which property is secure and further differentiated and workers are hired to produce goods, in ordinary employment, part of the labor value of the commodity goes to the employer in the form of profits or rent. The division of labor that creates profitability in manufacture produces capital for further investment, thereby increasing employment. The demand for goods and the realization that productive labor is a creative commodity, plus the available capital for investment, together create a climate favorable to industrialization and economic growth. Labor and capital therefore are interrelated, and as I suggested in Chapter 2 Smith implies that the security, differentiation, and protection of private property are necessary to generate capital for the growth of commerce, the specialization of labor, and the economic well-being of the laborers.[17]

## Smith's Labor Theory of Value

Smith notices that even primitive exchanges depend on the market. That is, what people want and what they can purchase or barter for create demand and so determine what labor needs to supply. Obviously, in an underpopulated or poor society such demands are less than in a more populated or opulent society. The exchange value of any item depends in part, therefore, on the extent of the market.

Smith, however, links with labor the "exchangeable value" of commodities, that is, the purchasing or bartering power of a commodity or what it can be exchanged for. Exchangeable value measures both the value of a commodity and the annual revenue of every society, that is, "the whole annual produce of its industry."[18] In writing about exchangeable value Smith distinguishes between productive and unproductive labor, connecting the former to exchangeable value. Productive labor is labor that adds value to what it works on—for example, to raw materials in manufacture, to land in agriculture—so that the resulting product is more valuable than the sum of the labor and the materials. Unproductive labor is service that does not add value and includes the work of servants, government officials, teachers, lawyers, doctors, ministers, and even military personnel. Because productive labor produces value, it also supports unproductive labor and thus is essential to an economy and economic growth. Relating productive labor to exchangeable value, Smith states:

> The value of any commodity, therefore, to the person who possesses it, and who means not to use or consume it himself, but to exchange it for other commodities, is equal to the quantity of labour which it enables him to purchase or command. Labour, therefore, is the real measure of the exchangeable value of all commodities.[19]

This is because the surplus that one produces determines what one can purchase or barter. Thus every exchangeable item has a labor value. To put the point in another way, if one exchanges item A for item B and makes a profit from that exchange, the resulting labor value will be the amount of labor that one can purchase with the proceeds, that is, the "commandable value" of the proceeds. Accordingly, only labor can produce value, because only labor can produce surpluses by its very productivity rather than its intrinsic value.[20] Exchangeable value also extends to rent and profits, but Smith argues that "labour measures the value not only of that part of price which resolves itself into labour, but of that which resolves itself into rent, and of that which resolves itself into profit.[21]

Smith comments further that what he calls the "real" value of labor is invariable. Because productivity is a commodity, it has a market or "nominal" price, which is distinct from the real price:

> [Labor's] real price may be said to consist in the quantity of the necessaries and conveniences of life which are given for it [its commandable value]; its nominal price, in the quantity of money. The labourer is rich or poor, is well or ill rewarded, in proportion to the real, not to the nominal price of his labour.[22]

Smith's aim, of course, is to find a universal standard of measure of economic value, and he finds labor productivity rather than use value or scarcity more amenable to establishing that standard:

> Equal quantities of labour, at all times and places, may be said to be of equal value to the labourer. . . . Labour alone, therefore, never varying in its own value, is alone the ultimate and real standard by which the value of all commodities can at all times and places be estimated and compared. It is their real price; money is their nominal price only.[23]

What Smith is claiming is that to the laborer the value of labor—or, more accurately, its productivity—is constant and universally so, although productivity obviously has different values to various employers and under variable market conditions. But one can measure the real value of any commodity by the amount of labor that it commands and the price of labor for producing that commodity. Labor has a "real" price in terms of its value to the laborers, an exchangeable value in what they can buy with its proceeds, and a market or nominal price in terms of what they pay. Moreover, at any one time, certain quantities of labor produce certain amounts of goods, and so labor is the best measure of the value of those goods. The invariability of the value of labor to the laborer establishes labor as "the only universal, as well as the only accurate measure of value, or the only standard by which we can compare the values of different commodities at all times and at all places."[24]

Smith's analysis raises some difficulties. By linking the real value of productivity to a psychological unit of measure as well as its commandable value, he is assuming an invariability of two highly volatile components of labor activity, an assumption that persists in some analyses of marginal productivity theory today. Also, by establishing a real labor value of a commodity and at the same time recognizing that no laborer is paid fully for his or her labor productivity, the rest going to rent and profits, Smith opens up an opportunity for Marx to use a labor theory of value to criticize the capitalist treatment of laborers.[25]

## The Freedom and Alienation of Labor

Smith finds the commodification, division, mechanization, and specialization of labor productivity to be the basis for economic freedom, growth, and economic well-being. Despite these positive conclusions, however, there appear to be inconsistencies in Smith's treatment of these phenomena. In Book I he writes about the benefits of the division of labor, but in Book V he states in regard to the specialization of labor:

> In the progress of the division of labour, the employment of the far greater part of those who live by labour, that is, of the great

> body of the people, comes to be confined to a few very simple operations; frequently to one or two. But the understandings of the greater part of men are necessarily formed by their ordinary employments. The man whose whole life is spent in performing a few simple operations, of which the effects too are, perhaps always the same, or very nearly the same, has no occasion to exert his understanding, or to exercise his invention in finding out expedients for removing difficulties which never occur. He naturally loses, therefore, the habit of such exertion, and generally becomes as stupid and ignorant as it is possible for a human creature to become. . . . His dexterity at his own particular trade seems, in this manner, to be acquired at the expence of his intellectual, social, and martial virtues.[26]

Many commentators contend that Smith contradicts himself between Book I and Book V. The division of labor, so important to the development of commerce, creates worker sloth and degeneration as the economy advances and the specialization of labor narrows.[27] Although Smith does not talk about the alienation of the laborer as a result of this deterioration, some readers find in these passages the roots of Marx's theory. Although the division and specialization of labor are necessary for industrialization, the treatment of labor as a commodity measured simply on productivity and efficiency and separate from laborers who produce that work gives laborers a sense of estrangement from their work and thus a sense of isolation from themselves and from fellow workers who are measured similarly.

There may be, then, a paradox in Smith's theory of labor and economic progress. He claims that a society makes economic progress in the sense of providing economic liberty and a decent standard of living for laborers as well as others, only when consumer demands that cannot be met create a climate for the industrialization and commodification of productivity that frees laborers from identifying with their productivity.[28] At the same time, Smith argues, it is only with the accumulation of capital and thus property inequalities—which he admits require the necessity of government "as a defence of the rich against the poor"—that the stage of commerce, real independence of laborers, and economic well-being can be achieved. That is, labor freedom and well-being are accomplished only in the context of such inequalities.[29] Accompanying this is a sense of powerlessness, because one's work is controlled and measured by one's employer. Although Smith appears not to have solved the problems of alienation, it is sometimes said that he did see these difficulties, which he found inherent in an industrialized free-enterprise economy such as he proposed.[30]

Smith obviously influenced Marx's theory of the alienation of labor. Smith's early premonitions about the negative effects of specialization triggered Marx's reflections, reflections that are taken seriously in the treatment of labor in most industrialized nations today.[31] Although Smith does not use the term *alienation*, it is evident that he deals with the issue. But had their historic roles been reversed, I believe, Smith might have seen himself as presenting an alternative to Marx's theory of alienation. Let us see why this is so.

In a preindustrial economy without the valuing and specialization of labor, workers are, or become, lazy because they are not in control of their own labor; their contribution is not measured; and their economic well-being cannot increase in any significant way. Similarly, in the age of commerce, a worker's accrued slothfulness in an advanced stage of specialization can be attributed to his "having no occasion to exert his understanding." Rather than being separated from their work, Smith implies that workers at this stage again slip back into identifying themselves with their work. He maintains, however, that this problem can be remedied with good public education and cultural diversions available to everyone. It is education, according to Smith, that enhances the lives and character of laborers,[32] the idea being that the progression of the mind helps one to judge one's work rather than to become identified with it. Whether education alone is enough to prevent this phenomenon of identification is surely questionable. But Smith's more important point is that it is the separation of the laborers from their productivity, not their identification with it, that creates the condition for liberty. Smith suggests in Book V that laborers who do not identify with their work are thereby more productive and innovative than are laborers who do, because in the latter case, they have "no occasion to . . . exercise their invention." Accordingly, then, Smith would argue that it is when workers identify with their work that they become self-estranged, because in those instances, they lose themselves to their work.[33]

One way to understand Smith's point, which is extrapolated from the text, is to borrow a term that he uses in the *TMS* and the *LJ*, the *impartial spectator*. In this context one can think of freed laborers as impartial spectators who can view their work disinterestedly and thus are capable of improving their productivity, bartering for higher wages, or even changing jobs. Smith also recognizes that laborers ordinarily do not act impartially:

> But though the interest of the labourer is strictly connected with that of the society, he is incapable either of comprehending that interest, or of understanding its connection with his own. His condition leaves him no time to receive the necessary information, and his education and habits are commonly such as to render him unfit to judge even though he was fully informed.[34]

But universal good education might be a means to develop habits or abilities of judgment or at least to enable the common person to be "less liable . . . to the delusions of enthusiasm and superstition."[35] Although Smith sees problems with lengthy overspecialization, he understands that because it converts abstract freedom into economic bargaining power, the division of labor is essential to freeing laborers from being treated as a commodity. Likewise, education and other diversions can free the laborers' minds from the torpor of work and from their identification with it and perhaps help them make disinterested judgments.

Concerning the alleged paradox between Books I and V, as Nathan Rosenberg points out, one can make too much out of the inventiveness of workers to which Smith refers in Book I. He is more likely to have thought that workers can be inventive in a narrow way having to do with their tasks, but elsewhere Smith states that inventions usually are created by those who are not specialized.[36]

Marx called Smith's notion of economic liberty "abstract liberty," because workers are never paid fully for their contribution.[37] Smith also recognized that fact. Yet because capital is necessary for industrialization, he does not thereby view the laborers' liberty as abstract. Without the development of private property and the subsequent industrialization, laborers are never at liberty. Or if they are "freemen," without a demand for productivity the laborers can never develop their "property" as an exchangeable commodity, nor will the economy achieve what Smith calls "universal opulence,"[38] which provides for the well-being of the laborers. Without the separation of the laborers from their productivity, they can neither barter for their productivity nor defend themselves against serfdom, because they cannot self-consciously conceive of both economic independence and the fact of the value of labor productivity. And without the specialization of labor, laborers will not be liberated from their reliance on agriculture and thus from the charity of their landlord, and economic growth will be curtailed.[39] It is true that industrialization and the division of labor do not guarantee the workers' well-being. But, Smith argues, both the distinction of the laborers from their productivity and the division of labor are the sources of economic liberty and economic well-being, and without both of the former, neither of the latter can be accomplished. So even though Smith would agree that industrialization separates the laborers from their productivity, he does not see this separation or alienation negatively. Rather, Smith would find the inseparability of the laborers from their productivity as a loss of freedom.

In response, although Marx recognizes industrialization as a necessary stage in economic development, at least some neo-Marxists, and probably Marx himself, would argue that the laborers, their labor, and their productivity that results from their labor are inalienably linked and thus inseparable. Even when a laborer voluntarily rents or sells his or her productivity, the commodification of labor productivity is a form of slavery in which the laborer barters away control of his or her productivity and thus gives up his or her freedom. It is only when productivity is decommodified and returned to the laborer that he or she is truly free.[40] On the other hand, Smith thinks that there is a logical separability between the laborers and their productivity that is a result of their labor and that without that separation, the laborers cannot be conscious of or in control of their labor and cannot thus be independent. Smith acknowledges that this independence is a form of abstract liberty until or unless a commercial economy provides the means to transform that independence into concrete well-being through the specialization of labor and the subsequent industrialization of the economy, but an identification of labor and labor productivity precludes even abstract liberty.

Smith develops no conception of a communal society without property distinctions that could generate capital for investment from its productive labor. Even though he views the social passions as coequal with the selfish ones, he questions the dominance of the former in economic affairs and wonders whether either laborers or managers would be willing to contribute efficiently and productively to the common good without personal remuneration. In addition, I speculate, Smith would argue that no form of cooperative economic enterprise could generate economic growth unless those participating in this enterprise saw themselves as independent and in control of their work. Otherwise, the identification of laborers and managers with their work and with what they produce would lead to psychological and thus productive decline and, eventually, to the stagnation of the economy.

## Economic Liberty of Laborers in a Society

Even if laborers are psychologically separated from their productivity and are part of a developing industrial process, they still might be restricted by societal policies and practices. Part of the purpose of the *WN* is to take issue with the economic practices and regulations extant in the eighteenth-century British economy of Smith's day, practices and regulations that Smith calls *mercantilism*. His attack on mercantilism is not merely directed toward its trade policies and regulatory monopolies; it is also directed toward what he regards as the philosophy of mercantilism, the view that economic activities can be orchestrated like chess pieces on the "board" of the economy.[41] Smith is particularly critical of the many mercantile practices that restricted the free movement and choices of laborers and manufacturers: "The policy of Europe, by obstructing the free circulation of labour and stock both from employment to employment, and from place to place, occasions in some cases a very inconvenient inequality in the whole of the advantages and disadvantages of their different employments."[42]

The condition of "perfect liberty," an ideal in the *WN*, was far from being realized in Smith's time. People were often told where they might or might not reside;[43] manufacturers and merchants were often prevented from engaging in the business of their choice; and some wages of labor were subject to restrictions.[44] The most repressive practice was that of apprenticeships. In Smith's time apprenticeships were encouraged as a means to develop skills, but they were generally of long duration with little or no pay. The apprentice was bound to his master and supported by his family. These programs, then, discriminated against those whose families could not afford to support them; they extended the time of apprenticeship far longer than necessary to learn a skill; and they unduly exploited the apprentice. Apprenticeships also did not extend to agriculture labor, which Smith often found to require more skill than many other trades did. Apprenticeships violated "the property which every man has in his own labour," according to Smith, because the apprentice neither was free to control his choice of work nor was paid for it.

In writing about working, Smith states his ideal:

> The whole of the advantages and disadvantages of the different employments of labour and stock must, in the same neighbourhood, be either perfectly equal or continually tending to equality. . . . This at least would be the case in a society . . . where there was perfect liberty, and where every man was perfectly free both to chuse what occupation he thought proper, and to change it as often as he thought proper.[45]

Smith realizes that even under optimal conditions of perfect liberty, equal advantages are seldom achieved because of changes in supply and demand, the introduction of new industries, and the involvement of many laborers in more than one job or profession.[46] Indeed, Smith stipulates five conditions that permit wage differentials: the hazards of the employment, the skill required or the technical difficulty of a job, constancy of employment, responsibilities of the job, and the probability of success.[47] Thus when Smith suggests that "the whole of the advantages and disadvantages of the different employments of labour and stock must . . . be either perfectly equal or continually tending to equality" the "equality" to which Smith refers is not literally that. Nevertheless, the liberty of choosing and changing employment, like other economic liberties, is an important part of the goal to be met in a just political economy.

But in another sense, is the economic liberty of laborers to choose their occupations merely abstract liberty? The liberty of laborers to choose and change occupations along with the specialization of labor, Smith believes, provides the initiative for economic growth. Yet wage earners, who must live solely from their wages, have little leverage in bargaining because they neither control employment nor have any reserves should they lose their jobs. Smith is therefore one of the early advocates of high wages, noticing that "where wages are high, accordingly, we shall always find the workmen more active, diligent, and expeditious, than where they are low."[48] Unfortunately, most employers do not view wage increases from this enlightened perspective. Whereas collusion among employers for keeping wages low was common and effective in Smith's time, collusion among workers to refuse to work was usually both fruitless and illegal. The alleged free choices of laborers, considered in isolation, might thus be thought of as abstract because their choices are controlled by the power of employers and the state of the economy. Thus, even if one can make sense of what appears to be Smith's resolution of the problem of the alienation of labor in light of the importance of the independence and division of labor for economic freedom and well-being, how is one to treat laborers fairly in a political economy in which inequalities of property are the "engine" that runs the infusion of investment capital and the power that controls labor?

Smith has three responses to this problem. First, as we found in Chapter 4, he believes that in the absence of regulations that favor one industry or group of business people, competition ordinarily exerts enough pressure on individuals or industries to preclude the development of monopolies or power bases that can accrue economic advantages over extended periods of time. Similarly, in a situation of economic growth and free trade in which there is not overpopulation, competition among employers for good workers raises wages and increases the bargaining power and well-being of even the lowest wage earner. In these circumstances, the demand for labor "naturally" increases wages, and so Smith concludes that the abolition of mercantile regulations and the market demand for workers will lead to the improvement of the laborers' standard of living.[49]

Second, Smith—who in general opposes regulation—remarks that although most laws pertaining to wages favor the employer and are usually unfair, those few that favor the worker, particularly those that require the worker to be paid in money rather than goods, are fair. He states: "When a regulation, therefore, is in favour of the workmen, it is always just and equitable; but it is sometimes otherwise when in favour of the masters."[50] The reason is that such regulations, ordinarily created by "masters" or employers, impose no hardships on employers, whereas without a law on their side, workers have no redress for their grievances. Smith, however, is not in favor of any laws that regulate wage rates, because these punish the able worker while overpaying the indolent one.[51]

Third, Smith argues that in a free society a proper framework of commutative justice should protect every person from harms, from the violation of contracts, and from unfair activities. Whether laws of justice are enough in a competitive commercial society to provide laborers with equal opportunities and protections against injustice depends on how one interprets the scope of the adjudication of fairness in economic affairs. Smith declares that this scope should be limited, but in various places in the *WN* he cites exceptions to this generalization, for example, when he finds that the law must enforce the payment of wages in money, when it should abolish apprenticeships, when it must regulate some banking practices, when it must enforce fire wall codes,[52] and when it must even restrict interest rates.[53] Whether Smith would envision extending such exceptions to other areas of the treatment of labor is a matter of speculation but not without precedent in the *WN*. He does contend in the *LJ* that property is an adventitious but perfect right requiring that violations be restituted and demanding the necessity of government in order to be protected. So it should follow that the development of labor productivity as a commodity should bring with it property rights offering civil protection that parallels the rights and protections regarding other forms of property. Smith alludes to this conclusion when he says that laborers must be at liberty to choose and change their occupations and when he finds laws protecting fair treatment to workers to be good laws.[54] But unlike his treatment of other forms of property, Smith does not develop parallel arguments for protecting productivity as a perfect right.

With due respect to a more traditional reading of Smith, there is little evidence to support the claim that Smith is a positive advocate of employee rights. He is not. In fact, he is often taken to task for his economic analysis of labor and even slavery. For example, he finds poor wages unproductive as well as unjust and points out that slavery is usually economically unprofitable as well as in violation of human rights. So Viner writes reprovingly:

> Smith, however, in his general treatment of the market, although often not when he is dealing with particular cases, writes as if he accepts as realistic the same psychological assumptions when he is considering the relationship of master and servant, landlord and tenant-farmer, employer and employee, as when he is discussing foreign trade.[55]

Viner is reading Smith correctly, but there is good reason for Smith's apparent coldheartedness. Employers, landholders, and slave owners have little in the way of fellow feelings for their employees or chattels. Moreover, as I stated in Chapter 3, neither pity nor benevolence guarantees the protection of rights or fair treatment. So laws of justice accommodate the impersonality of the treatment of workers while protecting their rights, and the link of poor treatment of slaves and workers to economic inefficiency strengthens Smith's arguments in regard to employees.

Smith's optimal solution is not to improve working conditions, guarantee jobs or minimum wages, or involve workers in any form of participatory management. Rather, he advocates free public education so that laborers can become less parochial in their outlook and can have other diversions. Nevertheless, Smith is appalled at employer abuses in eighteenth-century Britain, which he finds both unjust and economically inappropriate.[56] But except in the area of wages, in which he advocates the payment of a decent wage in money rather than in goods, government action is a last resort. Only when social and self-regulatory measures fail would Smith favor legislation to encourage fair play and fair treatment of laborers. Still, he maintains that laborers are better off in a "system of natural liberty" in which competition discourages the formation of monopolies or the setting of wage rates, because the laborers are better able to barter for their wages, and in the long run, their advantages will "tend to equality."

## Conclusion

Although Smith finds the commodification, division, and mechanization of labor to be the cornerstone of an ideal political economy, like the invisible hand, a number of conditions must be met in order for these phenomena to succeed in contributing to concrete liberty, economic growth, and "universal opulence." First, until productivity, the "fruit" of labor, is valued and recognized as an exchangeable commodity that also can create value, the division of labor cannot be a significant concept in a political economy. It is only when the productivity of labor is conceptualized as a commodity and as distinct from the laborer and when the productivity of labor is recognized as a measurable and expanding phenomenon that economic growth can take place and laborers are truly free. This usually occurs only when, second, there is an unsatisfied demand for goods that can be met only by expanding productivity with the specialization of labor. These two conditions are necessary but not sufficient, however. Third, there must be at least a modicum of capital for the initial industrial investment, in order to organize that specialization and mechanization. Fourth, as I suggested and shall argue in more detail in Chapter 6, regulations that favor one industry over another or that control the labor market prevent the free movement of and competition for labor, and thus restrict economic growth. Fifth, government must provide universal public education. Sixth, there must be opportunity for a market to expand, either within a nation or in international trade. Otherwise an industrial economy will become static or even decline. Even in a stagnant economic situation in which overpopulation, an abundance or lack of capital, or high wages produce a declining economy, Smith believes that the low prices of commodities produced by cheap labor would place that country in an advantageous trade position with its neighbors, thereby eventually improving the situation.[57] It is only when a society cuts itself off from trade that it becomes a static or declining economy. Finally, as we learned in previous chapters, an economy can flourish only in the context of justice, law, and order.

Smith is properly called an optimist in regard to the future of a free commercial economy, and an aristocrat in regard to the abilities of ordinary people to grasp what is in the public interest and thus govern themselves. He truly believes that what he calls a system of natural liberty holds better prospects for workers than does any other system. By freeing the workers from their productivity, they can become truly independent and not merely abstractly so. But such liberty is the liberty of economic activity and the liberty to exercise property rights, not the freedom of democratic choice,

Smith contends, because most of us have neither talent for nor insight into political affairs.[58] At one point Smith writes that the philosopher and the laborer are by nature equal, but he acknowledges that by birth, education, wealth, and luck they turn out differently.[59] Although labor productivity is the universal standard of economic value, the laborers almost never achieve economic equality with their employers. For Smith this is all right, because the laborers achieve economic liberty, opportunity, well-being, and equal advantages within their own milieu, achievements possible only in a commercial society based on private property. In response to Marx, Smith would see the identification of the laborers with their labor not as progress but, rather, as a return to a precommercial stage of serfdom in which laborers are exempted from personal freedom by being identified with their labor. Because the laborers are never paid fully for their productive contribution, as both Smith and Marx admit, it is more propitious to be in command of one's own labor productivity as a commodity for which one can bargain for the best exchange. It is true that Smith did not envision the evils for factory workers of the Industrial Revolution, nor did he predict the kind of negative alienation that, since Marx, we have regarded as a serious problem. One can extrapolate from what Smith writes in the *WN* to speculate that he would have been highly critical of industry's treatment of laborers in the nineteenth century. But his solution would not have been the dissolution of free enterprise, for he would have seen this as a loss of freedom and economic well-being for workers as well as for those with property. His solution to industrial abuses would have been regulation and universal education, for he would have found industrial workers alienated not by their separation from their productivity but by their having reunited with it.

# 6

# An Ideal Political Economy

> The *Wealth of Nations* has been represented in our century as one of the foundation stones of economic liberalism, free enterprise or competitive capitalism.[1]

The *Wealth of Nations* was written for a number of reasons, not the least of which is Smith's stated purpose in the title, to inquire into the "Nature and Causes of the Wealth of Nations." One of the explicit targets of Smith's inquiry is the eighteenth-century British economic philosophy and practice of mercantilism, and indeed, as we shall see, Smith devotes much of Book IV to an attack on what he finds to be questionable policies that do not contribute to what he claims to be a workable economic system. At the same time he presents an alternative to mercantilism—his depiction of a viable political economy that more satisfactorily answers his objectives: "First, to provide a plentiful revenue or subsistence for the people, or more properly to enable them to provide such a revenue or subsistence for themselves; and secondly, to supply the state or commonwealth with a revenue sufficient for the publick services."[2] But the objectives of a political economy are not merely economic or utilitarian, for Smith also states that a political economy should embody "the natural system of perfect liberty and justice."[3] That is, ideally it should be in harmony with the natural order.[4]

In this final chapter I shall present what Smith finds to be a viable political economy and analyze its relationship to the ideal he also sets forth. I shall examine his criticisms of mercantilism as they relate to what he finds to be negative elements in a political economy to see how his own views are both a remedy for mercantilism and a full-fledged theory of political economy. The chapter will conclude with a consideration of the status of Smith's political economy. If the *WN* presents the framework for a political economy, is what Smith presents a utopian ideal, or did he think that the model could actually be achieved? Is he optimistic about the future of an unregulated economy, or does he see its pitfalls, pitfalls that may portend its doom? Finally, is Smith truly the "father" of mature democratic capitalism? Is the *WN* truly the "foundation . . . of economic liberalism, free enterprise or competitive capitalism," or are Smith's views those of an eighteenth-century thinker that laid the groundwork for, but cannot be translated into, what we now take to be a free-enterprise system?

## Mercantilism

Beginning in the fifteenth century in Britain and the European continent, often conflicting laws and practices developed that were designed to regulate the economy of the day and preserve, if not increase, economic wealth and welfare. These practices, called *mercantilist* practices, dominated the British economy in the eighteenth century and are the subject of analysis, discussion, and criticism in the *WN*. Smith refers to mercantilism as a body of economic theory and a system of political economy. Historians of economic theory, however, disagree about his characterization of mercantilism as a system. Joseph Schumpeter, for example, refers to mercantilism as "that imaginary entity, the mercantilist system."[5] Mercantilism was probably less a philosophy or a theory and more of a hodgepodge of often contradictory economic policies, regulations, and practices. Nevertheless, Smith organizes these regulations and practices into a body of theory that he then systematically attacks.

According to Smith, the theory of mercantilism is based on four major theses. First, he contends, the mercantilists held that "wealth consists in money, or in gold and silver."[6] Even land and natural resources were thought to be less valuable than money and gold, because the mercantilists could not envision that one could use land or natural resources to increase wealth. Second and more importantly, there is only so much wealth in the world; that is, the amount of wealth in the world is static, and so the notion of real economic growth was an alien one. Third, in order for a country to preserve and increase its wealth, it must protect its own assets and somehow annex wealth from another country. Fourth, therefore, "the encouragement of exportation, and the discouragement of importation are the two great engines by which the mercantile system proposes to enrich every country."[7] Economic growth is to be achieved by sacrificing one sector of the economy for another, by propitious trade agreements in which one achieves a net gain of money or gold, or by war. Thus in Britain a complex set of laws was created that controlled the use of gold and money and regulated economic exchanges within the British economy and with other nations. The purposes of these regulations were to control the economy and its wealth and to monitor individual economic activities in order to eliminate those that decreased wealth or were harmful to the public welfare.

As a result of this economic philosophy, in England and in other "modern" countries of the eighteenth century, there were innumerable restrictions on trade and exchanges both within that country and on trade with other countries. England even rewarded "bounties" to manufacturers and individuals to encourage the exportation of certain goods. In particular, most mercantile regulations favored the export of labor-intensive goods, whereas the immigration into the country of skilled labor was discouraged or prohibited, as jobs were scarce. Import trade was permitted when there was a quid pro quo arrangement with the exporting country, and in theory importation was to be restricted to essential goods and raw materials not available in the country. The ideal, of course, was to maximize exports and minimize imports. The wealth of the country had to be conserved, for import trade would deplete the wealth of a nation, and conversely, exports would create a net gain in wealth. In fact, these restrictions and bounties, Smith points out in the *WN*, favored some trades and manufacturers over others, created monopolies, and caused higher prices for "home" products.[8]

In Book IV of the *WN*, Smith, preoccupied with the problems of mercantilism, conducts a three-pronged attack on this economic theory. First, he points out that the restrictions and regulations imposed in the name of mercantilism, including those that restrict free trade, are bad for the economy. Nonregulation, however, creates a more favorable climate for economic well-being. Second, embedded in these arguments is the claim that mercantilist economic policies violate natural liberty and equality of treatment and therefore are unjust. Third, Smith criticizes the philosophy of mercantilism by presenting a theory that refutes the idea that wealth is static.

In questioning the mercantilist definition of wealth, Smith observes that wealth is not identified merely with money, gold, and silver. Money is solely an "instrument of commerce,"[9] and gold and silver, said to be more durable, are not necessary for either war or trade. In addition, "the importation of gold and silver is not the principal, much less the sole benefit which a nation derives from its foreign trade."[10] Rather, the primary beneficiaries of such importation are the gold and silver merchants themselves. Gold and silver are commodities like any others whose values fluctuate and so should be treated as such.

On the issue of foreign trade, Smith contends that the restrictions on international trade in the name of mercantilism have disastrous effects on the economy. One of the early defenders of free international trade, Smith explains at length that such trade allows a nation to export surplus goods and import competing goods, thereby lowering consumer prices. Even more desirable, free trade allows a manufacturer to import raw materials not available in the home country and thus to expand the amount of manufacturing goods available for export. Conversely, trade restraints create higher prices, restrict the expansion and diversity of manufacturer and thus jobs, and permit certain monopolies to develop of those goods not allowed to be imported. About these monopolies Smith writes:

> To give the monopoly of the home-market to the produce of domestick industry, in any particular art or manufacture, is in some measure to direct private people in what manner they ought to employ their capitals, and must, in almost all cases, be either a useless or a harmful regulation.[11]

The common practice of rewarding certain manufacturers or agriculturists for exporting goods, too, results in unfair preferential treatment.[12]

Mercantile restrictions not only harm the economy, but they also restrict "natural liberty," as one's economic life is controlled by regulations. These restrictions and the monopolies they protect are also unjust. Speaking against the restrictions of the wool trade that prohibit both the export of wool and the import of woolen cloth, Smith asserts: "To hurt in any degree the interest of any one order of citizens, for no other purpose but to promote that of some other, is evidently contrary to that justice and equality of treatment which the sovereign owes to all the different orders of his subjects."[13]

Many of Smith's criticisms of regulations and governmental restraints in the *WN* are directed against mercantilism as a repressive and inefficient system of political economy. What is interesting and controversial is that many of his attacks are against business people and business practices that operate and thrive in a mercantile system that encourages monopolies and favors certain interests. Indeed, in at least one passage Smith claims that merchants and manufacturers are the "principal architects" of the mercantile system.[14] How, then, could he imagine that these same merchants and manufacturers would be less conspiratory in an unregulated climate?[15]

Smith may have exaggerated the role of the balance of trade in mercantilist practices, as most mercantilists did not feel that a balance of trade was the sole means of creating wealth. Most mercantilists, at least in the eighteenth century, also distinguished money from wealth. What was less clear to most of them was how capital, land, agriculture, and the division of labor could produce economic growth.[16] In response, it was Smith who made famous an idea originating with the French physiocrats with whose writings Smith was familiar: Wealth can be created.

The seventeenth- and eighteenth-century physiocrats, led by François Quesnay, argued that the basis for any social order was its economic order. The system of market exchanges is subject to laws independent of human society, "discoverable by reason," and these laws shape the social order. Wealth comes from productive labor and so can be created by increasing productivity. The physiocrats, however, unlike Smith, thought that productivity could be increased primarily from the output of agriculture, but they did not imagine that productivity could be increased substantially in manufacturing. Furthermore, although they favored what they called a "laissez-faire" economy with few economic sanctions and free trade, they also projected a positive economic role for government in controlling agriculture output and maintaining price supports for farm products.[17]

In the *WN*, while borrowing some of their ideas, Smith is critical of the physiocrats because "the capital error of this system . . . seems to lie in its representing the class of artificers, manufacturers and merchants, as altogether barren and unproductive."[18] Questioning the claim that wealth is static and expanding on the theses of the physiocrats, Smith maintains that wealth is not determined merely by the amount of gold in a country but, rather, by its resources, land, and labor and, more importantly, through the productive use of these resources and labor. Real economic growth cannot be achieved by conserving or importing gold or by means of restrictive trade policies. Instead, wealth is increased by expanding exchangeable value, "the power of purchasing other goods which the possession of that object conveys."[19] Productivity is enhanced, as we all well know, through the efficient investment of capital and the division and mechanization of labor in both agriculture and manufacture. Improving farm yields and raising the productivity of capital and labor in manufacture—thereby increasing goods, services, wages, and, eventually, capital—enlarges the real wealth of a nation.

The fact that wealth can be created through productivity, according to Smith, is not enough to achieve an optimal growth economy. The inefficiencies, favoritism, and economic inequalities that result from mercantile practices convinced Smith that in general, systematic economic regulations are both less efficient than their absence and unfair. Wealth is increased and economic growth is encouraged, Smith argues, by removing regulations that favor one group over another, that create or protect monopolies, that force or prevent certain sorts of manufacture or economic interchanges, that restrict labor, or that in some way prohibit someone from entering into an economic transaction.[20] Economic stagnation is prevented by permitting free internal and foreign trade. Scattered throughout the *WN* are details of at least eight reforms that Smith had in mind for the British economy, all of which I have cited. They include the free movement of labor—the opportunity to live where one wishes, a free choice of occupations, reform in apprenticeships, free trade and transfer of land, the abolishment of special privileges for joint-stock companies, and the ending of restrictions on internal and foreign trade. Although Smith does not lay out these reforms systematically in any one chapter in the *WN*, he implies that an economy that accomplishes these reforms is more efficient, more likely to achieve economic growth, and closer to the natural order of perfect liberty and jurisprudence.[21]

But what about the "principal architects" of the mercantile system, the merchants and manufacturers? Despite his skepticism about some business practices, Smith is convinced, as I argued earlier, that freed from restraints, competition among businesses will be such that monopolistic and unfair business activities will be at a minimum.[22] In other words, the principal architects and benefactors of a mercantile system will find themselves both freer to conduct business and more restrained—not by regulations and protections but by the equally free business activities of their competitors, from whom they are no longer protected.

## The Scope of Economic Freedom in Smith's Political Economy

Having attacked the philosophy of mercantilism and what he finds to be the repression and favoritism of mercantile policies and a mercantile system, Smith proposes an economy in which individual choice is the norm for economic motivation and action. About such an economy Smith writes: "In general, if any branch of trade, or any division of labour, be advantageous to the publick, the freer and more general the competition, it will always be the more so."[23] Thus although the *WN* does not spell out in an orderly fashion what Smith concludes to be an ideal economy, it is obvious that he is a strong proponent of "natural" economic liberty. Therefore, his critique of mercantilism is often cited as an endorsement of a laissez-faire, unregulated "night watchman" economy. As I stated in Chapter 4, Smith's eighteenth-century moderate collectivism, the assumption that a political economy is embedded in an institutional, social, and religious framework, and his focus on economic but not political liberty eliminate that conclusion. Yet one must be careful not to underestimate or overestimate the extent to which Smith defends economic liberty. As he says in the *TMS* and the *WN*, each of us, naturally, is better able to take care of ourselves than others can, and a "system of natural liberty" that both removes special privileges and encourages independent self-improvement best emulates the ideal of caring for oneself. As Smith contends in the *LJ* and the *WN*, the best kind of economy is one of free commerce in which "the law ought always to trust people with the care of their own interest."[24]

Smith justifies a political economy that emulates economic liberty on three grounds: Such an economic system best reflects the natural order; it is less unjust; and it is most efficient. The elements of the natural order—caring for oneself, liberty, natural rights, and natural jurisprudence—that justify a nonrestrictive economy are, at the same time and from a practical point of view, the "engine" that runs it:

> The natural effort of every individual to better his own condition, when suffered to exert itself with freedom and security [of protection by laws of justice], is so powerful a principle, that it is alone, and without any assistance, not only capable of carrying on the society to wealth and prosperity, but of surmounting a hundred impertinent obstructions with which the folly of human laws too often incumbers its operations.[25]

Therefore, under the protection of laws of commutative justice, the free flow of economic exchanges is more likely to produce economic prosperity and be in the public interest as well.

An ideal economy is one that improves the productivity of agriculture and, more importantly, of manufacture. Because it must at least implicitly recognize the freedom of workers when valuing their productivity as an alienable, exchangeable commodity, such an economy also is the most likely to guarantee natural liberty to everyone. Thus Smith's ideal economy is one of universal economic liberty, free commerce, and industrialization, although it is unlikely that he recognized the extent to which a nation could become industrialized as subsequently Britain and, later, America did.

There are, however, at least two extrapolative errors often committed in Smith's name. First, he is often called the "father of democratic capitalism." Second, as I pointed out in Chapter 4, Smith is sometimes depicted as a nineteenth-century "night watchman" individualist. It seems logical to identify natural liberty with political freedom and an unregulated economy with laissez-faire capitalism. But as we saw in Chapter 2, in examining Smith's historical description of the progress of civil society, he has little confidence in democracy as a form of government except in the most primitive societies in which property and property rights are not at stake. This is not because he thought that the poor would rise up against the rich, although they might encroach on another's property. Nor is it merely because of the tradition of monarchical authority, the respect of which helps keep peace and order. Instead, Smith says in this regard:

> There are two principles which induce men to enter into a civil society, . . . the principles of authority and utility. . . . Age and a long possession of power have also a tendency to strengthen authority. . . . But superior wealth still more than any of these qualities contributes to conferr authority. This proceeds not from any dependance that the poor have upon the rich, for in general the poor are independent, and support themselves by their labour, yet tho' they expect no benefit from them [the wealthy] they have a strong propensity to pay them respect.[26]

Thus he understands the value of the tradition of the authority of wealth in eliciting obedience and thus maintaining peace and order. He also finds that the best form of monarchy is one in which the parliamentary and judicial powers are separate, in order to prevent tyranny or absolutism. Parliamentary powers, however, are limited to landholders and the aristocracy.

Smith's greatest apprehension concerns the decision-making abilities of the common person: "Though the interest of the labourer is strictly connected with that of society, he is incapable either of comprehending that interest, or of understanding its connection with his own."[27] And about landholders he remarks: "That indolence, which is the natural effect of the ease and security of their situation, renders them too often, not only ignorant, but incapable of that application of mind which is necessary in order to foresee and understand the consequences of any publick regulation."[28] Only merchants and manufacturers have some knowledge of the public interest, but even "the interest of the dealers . . . is always in some respects different from, and even opposite to, that of the publick."[29] It would be disastrous to civil society and economic progress to allow democratic decision making to become the rule of law, as our private interests are generally too parochially directed. Therefore, for Smith, "natural or perfect liberty" is economic liberty, the freedom to deal with one's personal affairs and to improve one's own condition, tasks that each of us is best capable of doing, but natural liberty does not include the idea of democracy except in a narrowly restrictive legislative form.

Smith is surely the strongest eighteenth-century proponent of what we now call private free enterprise. But he is most aptly named the "grandfather of modern capitalism." Although it is obvious that Smith is critical of mercantilism, what he has in mind is not an advanced corporate/industrial system. His ideal economy, rather, is that of "mom and pop" stores, small manufacturers, large landholders with tenant farmers, and entrepreneurial merchants. Smith is highly critical of "joint-stock companies" or corporations:

> The trade of a joint stock company is always managed by a court of directors. The court, indeed, is frequently subject, in many respects, to the controul of a general court of proprietors. But the greater part of those proprietors seldom pretend to understand any thing of the business of the company. . . . The directors of such companies, however, being the managers rather of other people's money than of their own, it cannot well be expected, that they should watch over it with the same anxious vigilance with which the partners in a private copartnery frequently watch over their own. . . . Negligence and profusion, therefore, must always prevail, more or less, in the management of the affairs of such a company.[30]

Smith's animadversions on the East India Company—perhaps one of the first multinational corporations—demonstrate his fear in that regard.

Concerning Smith's alleged economic individualism, he argues persuasively against government regulation, but it is oversimplistic to conclude that "the ideal system is one of fragmented economic power in which self-interest leads participants to cooperate with each other in transactions which yield mutual gain."[31] There are a number of reasons that this is a questionable reading. First, interests in the form of non-tuism and cooperation are two sides of the selfish and social passions that work hand in hand in economic (and other) activities. Even though it is in our interest to cooperate, cooperation is also part of our natural social passions, not merely an outcome of self-interest. Second, as Chapter 4 demonstrated, Smith argues less in favor of fragmented economic power and more strenuously for fair competition among economic powers, however strong and organized they are, because, he believes, fair competition "levels" power differences among competitors.

Third, the notion of a fragmented economic system gives the impression that Smith favors an economy in which there are no restraints whatsoever. This is erroneous. Although it is true that he thinks that an unregulated economy is fairer and closer to the natural order, because no one is favored or protected, it is the economic affairs of a society, not the polity itself, that should be allowed to flourish in an atmosphere of perfect liberty. Moreover, as I stated in Chapter 4, even the free market cannot function efficiently or fairly except within the safeguards of a well-defined system of justice.

Fourth, Smith finds positive roles for government in education and public works. Smith criticizes governmental interferences in private lives, for example, restrictions on where one can live or work, apprenticeships, or religious constraints, and he describes the duties of government in very simple terms:

> According to the system of natural liberty, the sovereign has only three duties to attend to . . . : first, the duty of protecting the society from the violence and invasion of other independent societies; secondly, the duty of protecting, as far as possible, every member of the society from the injustice or oppression of every other member of it, or the duty of establishing an exact administration of justice; and, thirdly, the duty of erecting and maintaining certain publick works and certain publick institutions.[32]

The importance of this third duty of government is sometimes not given sufficient attention. An essential role of government, according to Smith, is

> erecting and maintaining those publick institutions and those publick works, which, though they may be in the highest degree advantageous to a great society, are, however, of such a nature that the profit could never repay the expence to any individual or small number of individuals, and which it, therefore, cannot be expected that any individual or small number of individuals should erect or maintain.[33]

Furthermore, while extolling free commerce and industrialization, Smith sees that some regulations of the economy are necessary. The role of government is limited, but sometimes in the *WN* Smith extends its activities to include interference into questionable commercial practices. For example, he finds that some regulations may be necessary to prevent unfair banking practices; fire codes may have to be enforced;[34] and interest rates may have to be regulated. Smith is not opposed to regulations that favor workers, as I mentioned in Chapter 5, and government may also have the task of preventing or breaking up monopolies.[35] In being critical of monopolies and collusion among dealers, manufacturers, or merchants, Smith argues that one of the functions of government is to "protect . . . every member of society from the injustice . . . of every other member." Smith writes that "it can never be the interest of the proprietors and cultivators to restrain or to discourage in any respect the industry of merchants, artificers and manufacturers."[36] In this passage he intimates that government might have a role in breaking up monopolies or collusive agreements among business people that interfere with competition or are not in the public interest, although he does not explain this role, perhaps because he is fearful of government regulation, as it usually favors one group over another. Or perhaps the reason is that he had not developed his theory of justice in which laws of justice, and therefore specific regulations against monopolies and collusion, would most clearly reflect the precepts of natural jurisprudence.

Therefore, although in principle Smith leans toward free economic exchanges, he would be skeptical about the viability of competition among large joint-stock companies; he would worry about their powers being separated from their responsibility; and he might not be inclined toward the nonregulation of secondary markets, for example, banks and stock markets. Finally, as I have reiterated a number of times, Smith concludes that no economy can exist without a strong foundation of justice buttressed by other social and political institutions that provide a background of law, order, and continuity of religious tradition and morality. These supports are necessary because we tend to translate Smith's idea of an unregulated economy into nineteenth-century liberalism or twentieth-century mature, democratic industrial terms. Smith, however, was firmly rooted in the eighteenth century.

## Smith's Relevance to Contemporary Economic Theory

If Smith were an eighteenth-century political economist who questioned the viability of a political democracy and whose concept of advanced capitalism was rudimentary at best, one would have to respond to Warren Gramm's conclusion that the *WN* should be considered primarily from a historical point of view as an eighteenth-century text written for the economy of that time. In a series of articles, Gramm argues that it is a mistake to translate or reinterpret Smith's ideal political economy and the concepts it embodies into an economic theory applicable to twentieth-century industrial or postindustrial corporate capitalism. What the *WN* has to say is historically important and influential, but its conclusions are most relevant in the framework of an eighteenth-century precorporate entrepreneurial economy in the infancy of the Industrial Revolution. As such, Gramm reasons, the arguments in the *WN* do not and cannot apply to twentieth-century economic theory or practice.

Gramm adds another kind of argument. Smith was writing during the eighteenth-century preindustrial, precapitalist era. The Industrial Revolution was in its infancy. The corporation as an industrial organization was scarcely in existence. Smith assumes that his political economy is part of a social system, the system existing in eighteenth-century England. "For Smith, the metaphor of the invisible hand calls attention to the paradoxical simultaneity of diversity (conflict) and community in individual interests"[37] in his era. The invisible hand works only in that context, a context that presupposes a number of elements no longer operative in today's markets. As we noticed in the "butcher, baker" example, Smith imagines economic exchanges in small towns in which community relationships present an obvious constraint. The butcher seeks approval of his fellow citizens, who are also his neighbors and friends. Today with large corporations and absentee ownership, community relationships are not as important and do not restrain economic activities to as great an extent.

Second, Smith's notion of natural liberty assumes open entry into markets. This is possible only because he envisions business as small business and, indeed, is highly critical of what he calls joint-stock companies. Smith early admits the problem with dividing ownership, and thus ownership responsibility, from management.[38] But in this century, the necessity for large amounts of capital preclude open entry into the market by small businesses.

In addition, according to Gramm, the invisible hand is to work within the constraints of a strongly religious community, in Smith's case the restraints of Protestant Calvinism, to which hard work and parsimony were of great value. This "social–humanist perspective" is peculiar to eighteenth-century preindustrial England, as both Robbins and Gramm acknowledge. But in most communities and in most businesses today, pluralism prevents religious constraints on economic activities. Thus, Gramm concludes, it is a mistake to transform Smith's economic concepts into contemporary economic theory, both because this transformation distorts what Smith meant and because the terms themselves are not relevant to twentieth-century economic life.[39]

Gramm is right that we have removed Adam Smith's economic theory from its eighteenth-century context. Beginning with Smith's death and continuing to the present day, some thinkers have reinterpreted his notions of self-interest, the division of labor, the invisible hand, and the limited role of government to suit their own terms. However, the deed was done early on, and for two centuries Smith has been read by some as an egoist or, at best, as believing that self-interest is the dominant motive in economic affairs. Because certain interpretations of Smith's theses have strongly influenced economic theory and practice—in particular, rational choice theory—interpretations that I challenged in the preceding chapters, it is important to make sure that we are reading Smith properly and to correct any errors, so that what Smith might have concluded about a modern free-enterprise economy is consistent with what he actually did propose.

## The Dark Side of the *Wealth of Nations*

Some commentators find Smith to be a utopian whose laissez-faire model is an ideal impossible to realize. Typical of this view is Gunnar Myrdal's statement: "A sunny optimism radiates from Smith's writing. He had no keen sense of social disharmonies, for interest conflicts. . . . On the whole, it is true to say that he was blind to social conflicts. The world is for him harmonious. Enlightened self-interest ultimately increases social happiness."[40] But as I have suggested, this interpretation belies much of what Smith worries about in the *WN*. He finds fault with the fate of labor when it is allocated to small repetitive tasks; he sees problems with collusion between merchants and manufacturers; he details countless situations in which business people act against the public interest; he knows that property ownership yields inequalities; and he recognizes that unfair disadvantages occur even under the condition of perfect liberty. Smith, then, is hardly a "sunny" utopian.

In Smith's documentation of the failures of capitalism, Robert Heilbroner finds what he calls a "dark side" to Smith's analysis. Heilbroner argues that Smith himself predicted the moral decay and decline of capitalism, that his aim for any society is the "natural progress toward improvement" of both liberty and material conditions. Yet paradoxes that result from perfect liberty show that Smith allegedly developed a pessimistic view of capitalism, finding that his ideal political economy would eventually create economic and moral decay, for both capitalists and laborers. According to Heilbroner, Smith did not see a solution for these paradoxes, but he did see the problems that advanced capitalism might create.[41]

Heilbroner interprets Smith's analysis of the four stages of the development of property and civil government as four progressive stages in the historical evolution of economic growth that describe a model for all economic evolution. Our natural desire to "better our condition" is such that "in favourable circumstances society both *will* and *should* pass through these stages in sequence,"[42] according to Heilbroner's reading of Smith. In an earlier chapter I questioned such an interpretation of Smith's so-called philosophy of history. But given that reading, Heilbroner then criticizes Smith for truncating the description of the historical process of economic development by concluding that the last stage is an age of commerce. Smith did not do justice to his own insights, according to Heilbroner, because he himself saw the eventual decline of an unregulated private property system.

It is true that although Smith is a strong advocate of economic growth and finds that the age of commerce—with freedom, industrialization, and specialization of labor—provides the conditions for such growth, he does not paint a completely optimistic picture of the future of a commercial society. For example, he sees perfect liberty as leading to the improvement of standards of living for the laborer and sees the demand for labor as raising wages and thus reducing capital and economic growth. In a nation where there is full employment and wages continue to rise because of the demand for labor, Smith predicts that profits will fall and that capital will be unavailable for reinvestment. He sees that in a prosperous society the population increases; the demand for higher wages reduces stocks; and eventually both wages and stock decline. Alternatively, if there is too much capital for reinvestment in proportion to the demand for goods, the competition for goods will lower profits. Smith does not detail these possibilities but merely implies their likely occurrence in the distant future. Nevertheless, in such cases economic growth halts, and the economy becomes stagnant or static and eventually declines.[43]

> In a country fully peopled in proportion to what either its territory could maintain or its stock employ, the competition for employment would necessarily be so great as to reduce the wages of labour to what was barely sufficient to keep up the number of labourers, and, the country being already fully peopled, that number could never be augmented. In a country fully stocked in proportion to all the business it had to transact. . . . [t]he competition, therefore, would everywhere be as great, and consequently the ordinary profit as low as possible.[44] [The] usual market rate of interest . . . would be so low as to render it impossible for any but the very wealthiest people to live upon the interest of their money. . . . It would be necessary that almost every man should be a man of business, or engage in some sort of trade.[45]

Certainly this is not a positive outlook toward economic and social happiness. Smith, however, reveals in the next paragraphs that in such a situation, low prices of commodities produced by such cheap labor would place that country in an advantageous trade position with its neighbors, thereby eventually improving that economic situation.[46]

Heilbroner points out that Smith sees a "dark side" on the microeconomic level as well. The division of labor, essential to industrial capitalism, also raises wages, but Smith also recognizes that the division of labor with the repetitiveness of its tasks also impairs the moral character and intelligence of the workers. Therefore, Heilbroner argues, we have the first recognition of what later came to be known as the alienation of labor resulting from industrialization.

Smith also predicted overpopulation problems. A well-paid laboring class produces more children, who expand the work force and thus contribute to a decline in wages. Smith noticed, too, that excess profits lead to both avarice and the moral decay of capitalists and even of the poor. Landholders, and even manufacturers and merchants, seldom take into account the public interest and often collude with one another to control wages or prices. Smith's ideal of perfect liberty, then, as Heilbroner sees it, when linked to a laissez-faire economy and economic growth, eventually leads to moral and economic decay.[47]

Heilbroner makes little of the fact that although Smith acknowledges these problems, he also proposes solutions. Smith thinks that a civil government that enforces laws of justice and develops public works can both ameliorate private activities not in the public interest and preserve the ideal of competition. One of the public works that government should undertake is the education of every citizen, for it is education, according to Smith, that will improve the plight of the laborers. And one must not forget that Smith sees the demand for goods and the division of labor as means to liberate human beings from their reliance on agriculture, to force laborers to become independent of the charity of their landlords, and, more importantly, to expedite the development of the laborers' liberty from their labor. This can be realized only through economic growth, thus the division of labor functions to improve both the liberty and the economic status of laborers.

Heilbroner interprets Smith as a pessimist because, he argues, Smith sees free enterprise both as the last stage in economic progress and as having its own built-in disintegration. According to Heilbroner, Smith predicts the decline of capitalism, yet he could not find solutions to these paradoxes. Heilbroner traces Smith's failure to find solutions to his inability to imagine a solution to population growth and the finite availability of capital. Moreover, Smith could not see beyond capitalism to envision further historical–economic progress, a fifth stage in historical–economic evolutionary development.

Contrary to Heilbroner, I believe that although Smith found problems and even paradoxes in his ideal political economy, he saw solutions to these paradoxes within the system he proposed. According to Smith, an unregulated commercial and industrial economy is an ideal system—better than mercantilism or other forms of regulated economy because of its basis in natural liberty and justice and its promise of economic growth. But the system and thus the invisible hand work best only when the five conditions I cited earlier—perfect liberty, internal self-restraint, coordination, laws of justice, and perfect competition—are operative. The proposed system, then, is an ideal, because perfect liberty is seldom achieved; not every individual is willing to restrain his or her self-interests; cooperation is not always in the public interest; laws of justice are not always complete or fairly enforced; and perfect competition is rarely achieved, because advantages are not usually distributed similarly among competing parties. The division of labor, essential to the economic well-being of an economy, reaches its ends only when in addition to the foregoing conditions, there is available capital for investment, an unsatisfied market demand for goods, and an opportunity for an expanding market. Smith's lists of failures of laissez-faire are not "acknowledged exceptions" but, rather, failures of the ideal to be realized. He is somewhat pessimistic because he finds these failures to exist, but he is not pessimistic about his ideal. For if these conditions could be achieved, a "natural harmony in the economic order" would be possible. Therefore, Smith is neither a utopian nor a pessimist about his ideal of a political economy.

One can only speculate what Smith's reaction would have been to Marx and also to Heilbroner's analysis. In Chapter 5 I suggested that Smith's theory of labor is an anticipatory response to Marx regarding alienation. It is fairly certain that Smith did not see economic progress as necessarily following a certain historical pattern. He might not have concluded that the age of commerce is the final state in economic development; indeed, nowhere does he write that it is. But he would find his model for a political economy to be the most workable one, given the conditions of history as he could foresee them. Because Smith had no trust in joint-stock companies, it is unlikely that he could have imagined proletariat ownership and management of an economy. He had no confidence in the political or economic judgment of most human beings, either singly or in collective organizations in which responsibility was separated from ownership. Just as joint-stock companies do not always keep in mind the best interests of owners or the public, so too, on a larger scale, a proletariat might produce an unjust and inefficient political economy.

## A Viable Political Economy

Heilbroner's critique highlights the fact that Smith is not a utopian. Jacob Viner, asserting that the *WN* is less utopian than the *TMS* is, compiled a comprehensive list of passages in the *WN* in which Smith documents problems resulting from perfect liberty and competition, many of which were discussed in earlier chapters. They include inequalities of employment advantages even in a system of perfect liberty; the dichotomy between high wages and profits and between profits and capital investment; the avarice and laziness of both the wealthy and the poor; the possibility of the degeneration of labor as a result of its specialization; the collusion between manufacturers and merchants to manipulate trade, raise prices, and keep wages low; and the general lack of foresight that neither considers public interests nor takes the initiative in building or repairing public works.[48] Smith admits that we are often greedy, indolent, or shortsighted. We often act against the public interest, particularly in economic affairs, by forming monopolies, making private price or trade agreements, consorting to pay low wages, and even conspiring to cheat our employers. These weaknesses of free enterprise can be traced to a lack of vision, imprudence in economic affairs, parochialism, collusion, an unequal distribution of advantages, or simply a failure of justice in the system itself. Viner argues that in these many instances "Smith acknowledged exceptions to the doctrine of a natural harmony in the economic order even when left to take its natural course."[49] That is, Smith cited these problems because he was a realist about the fallibilities of human nature, not because he questioned his ideal. Even though in the end Smith found his political economy to be the best system, according to Viner, Smith saw that his ideal could not be perfectly realized.

Nevertheless, according to Viner, in the spirit of antimercantilism, Smith finds the evils of nonregulation preferable to those of a corrupt government that favor the interests of only part of the population. The three functions of government, security, justice, and public works are adequate to prevent monopolies, preserve the justice of agreements, and enhance the lives of laborers by providing public education. For, as Viner stated:

> Even when Smith was prepared to admit that the system of natural liberty would not serve the public welfare with optimum effectiveness, he did not feel driven necessarily to the conclusion that government intervention was preferable to laissez faire. The evils of unrestrained selfishness might be better than the evils of incompetent and corrupt government.[30]

Smith also combines and overlaps descriptive and normative elements in his analysis of the political economy. And for all these difficulties that he describes, he also offers possible solutions. For example, although he sees the alienation of labor, he also finds its solution not merely in education but also in the freedom of the laborers from their productivity. Property rights guaranteed in a civil society may create harms and thus injustices because civil government protects the rich and the propertied against the propertyless. Yet the existence of a "level playing field" of competition disperses the power of the rich into productive capital investment. When productive labor is valued as a commodity, the propertyless achieve economic liberty and have better opportunities. Even though we are often shortsighted and greedy, a commercial economy is still a better option than mercantilism is because there is—also—equal opportunity for greed, collusion, and shortsightedness. That we are not perfect rational utility maximizers and that part of our shortsightedness is due to our social passions and interests convince Smith that for those of us who do not internalize a sense of justice, laws of justice are a necessary and adequate substitute. Because we have social as well as selfish passions, we understand and accept the enforcement of commutative justice even when it is not in our self-interest, although we might not, by ourselves, always engage in fair practices. Perhaps Smith is overoptimistic that "the evils of unrestrained selfishness" will not dominate economic affairs and that competition can control monopolistic tendencies, and perhaps he relies too much on the "natural order" of the market to adjudicate economic unfairness.

Smith presents an optimal scheme for a realizable political economy that balances private and public interests, that maximizes natural liberty and opportunity for laborers and the poor as well as the wealthy, while also increasing economic well-being. Because his political economy is not utopian, there are trade-offs that may not be acceptable to all economic thinkers, but for Smith such trade-offs are balanced by a generally improved climate of natural liberty and economic welfare. If one considers his political economy in its eighteenth-century economic political context and if one analyzes his scheme in its totality rather than abstracting self-interest, individualism, the invisible hand, property rights, the division of labor, or nonregulation as representative of his theses, then Smith's political economy becomes a viable possibility in that context, and he can be seen as an alternative to both Marxism and the radical individualism of the nineteenth-century social Darwinists.

# Conclusion

This book has engaged in a sustained argument that some contemporary social science and economic paradigms erroneously trace their roots to Adam Smith and, in particular, to the *Wealth of Nations*. Many of their tenets are anticipated by Smith's analysis of a political economy and are answered by his arguments, albeit in the context of an eighteenth-century preindustrial outlook.

Smith's analysis of human motivation is a direct attack on the forms of egoism familiar to him in the eighteenth century. Agreeing with his contemporaries that we are motivated by passions rather than by reason, Smith contends that the existence of the social passions precludes the conclusion that human beings are merely egoistic, because our passions are directed to others as well as to ourselves. Our interests, too, though largely parochial, are equally self-interested and other-directed. Moreover, each of us desires not merely to be approved by others but also what ought to be approved, and so our desires include the preference to emulate what a society takes to be valuable rather than only what we or others generally prefer. Therefore excellence, not preference, is the normative model for human behavior.

Some contemporary economists believe that at least when acting rationally, human beings are rational utility maximizers, a view whose origin is sometimes attributed to Smith. This thesis does not necessarily imply that each of us is narrowly self-interested, for rational choice theory allows for altruistic human beings who have genuine interests in others. It is merely that in economic exchanges we are non-tuists. Smith recognizes this non-tuism, but he also shows that any economic exchange entails cooperation as well as competition, the former deriving from our social passions. Smith's view also questions whether we act rationally, and he finds that most of the time we do not. Rather, our actions are based on a variety of passions and interests, and we seldom take a long-term view of maximizing our own advantages or the advantages of others. Smith finds this pervasive but not unadmirable.

Even when we make rational choices on the basis of alternatives—and Smith would find this to be a relatively rare occurrence—even in these "cool moments," Smith argues, our choices are normatively influenced by societal values, and we do not always maximize utility for either ourselves or our interests in others. Human beings are not often motivated by, or base their moral approbation on, utility. In the *TMS*, Smith states that utility is only one of the sources of moral approval, and utility is not the end of morality. The reason is that utility is a general term not uniquely applicable to morality, because our preferences as well as our moral judgments are not guided merely by utility and because we often appreciate utility for its own sake, for its "beauty" as well as for what it may produce. Smith would not only dispute the fact argued by some economists and social scientists that we are primarily self-interested but would also doubt that when we act rationally in economic affairs, we always seek to maximize utilities of one sort or another.

In the *WN*, it is evident that motives of economic actors are not merely utilitarian, because the desires for approval and cooperation with fellow business persons also are motives for economic exchanges, even when increasing the economic welfare of another is not. Nor does the term *utilitarianism* exhaustively describe Smith's political economy. The invisible hand—the outcome of market exchanges—may lead to economic well-being, but one can hardly say that this is its aim, as the invisible hand has no intentions. In the *TMS* it is justice, not utility, that is a virtue, in fact the basic virtue, and in the *WN* it is justice, not utility, that provides the underlying framework for any society, even a society that "subsists[s] . . . from a sense of its utility, without any mutual love or affection."[1] So Smith is not merely a utilitarian even in the *WN*.

The goals of a political economy, too, are mixed. Its stated aims are to provide revenues, to enable each of us to better our own condition, and to supply public services and public works. At the same time, Smith feels, the duties of the various branches of government are to protect citizens from external or internal harms and also to guarantee "equality of treatment," protect rights, and administer justice. Any desirable political economy emulates natural jurisprudence and thus does not have merely utilitarian goals.

Smith is clearly an advocate of personal economic liberty. But, I have argued, Smith is neither a methodological individualist nor an ontological collectivist. Bracketing the question of whether one can apply modern terminology to a historical period in which those categories were differently conceived—and surely a historian such as Warren Gramm would take me to task for that—Smith clearly does not claim that each of us is a "social atom."[2] His remark in the *TMS* that "it is thus that man . . . can subsist only in society," his arguments that principles of justice are necessary for any viable society, and his analysis of business activities as interchanges in a community of cooperation all deny the conclusion that Smith believed that "economic man, self-interested and fundamentally asocial . . . is the model that explains human motivation and action,"[3] even when we are in fierce economic competition with another individual or group of individuals. Smith also does not envision economic liberty as isolated from law and social order. So even though he is an ontological individualist, he is certainly not a methodological individualist, and his ideal of economic liberty cannot be translated into a form of radical individualism.

The predominant normative theme running throughout the *TMS*, *LJ*, and *WN* is justice. Justice is defined as natural jurisprudence in the *TMS* and *LJ*, the basic virtue, and the underpinning of any social and political order. In the *WN* Smith argues that laws of justice are the essential feature of any political economy, a feature that should emulate as closely as possible natural jurisprudence. Recall that Smith's principles of justice include 1) not harming another or oneself; (2) engaging in fair play or, alternatively, not engaging in activities that are unfair; and (3) not engaging in conduct that violates perfect rights. Perfect rights include the natural rights not to be harmed, the right to personal liberty and personal reputation, and the adventitious right to property.

Within the framework of justice Smith places natural liberty and adventitious but perfect rights to property, rights essential to a modern free-enterprise economy. He then uses this model as the basis for a commercial market economy in which economic liberty, private property, the freedom and commodification of labor, and justice all are conditions for a smoothly running commercial society. Also embedded in that idea of the market is the norm of the prudent economic actor who cooperates to the extent of enabling workable exchanges. Smith's ideal, then, is universal economic liberty within the context of law and commutative justice following as closely as possible the natural order. This ideal both provides the grounds for and is possible only in an industrialized market economy, an economy of free labor and equal economic opportunities but not of equal outcomes. In such an economy individuals are neither economic atoms nor organically unified but, rather, depend on and contribute to a background of social interaction to achieve their economic interests.

The notion of benevolence, so often called the crowning virtue of the *TMS*, is, as I have stated, not the only virtue or the basic one, even in Smith's moral psychology. A denial that Smith has a model of benevolent capitalism does not mean endorsing a view of economic man as asocial or egoistic or the idea of pure laissez-faire as the ideal. The fact that benevolence is a virtue peripheral to economic exchanges is neither surprising nor damaging to Smith's model. For an ideal political economy is governed by the principles of jurisprudence that apply to everyone, to all exchanges, and to them all equally. Thus fairness, not benevolence, is the cardinal virtue of individuals, economic exchanges, and a political economy itself.

In summary, Smith's description of human motivation is neither egoistic nor altruistic; his analysis of labor and the productivity of labor is essential to economic liberty; and his focus on economic liberty, adventitious property rights, and the market is in the context of laws of commutative justice. Despite the historic notoriety of the term invisible hand, Smith's philosophy of economics argues against the personification of the market and an overglorification of its intentions and control, while avoiding some of the pitfalls of rational choice theory. Smith's ideal economic actor is a person of goodwill, prudence, and self-restraint who operates both cooperatively and competitively in a social and economic milieu based on a foundation of morality, law, and justice. Thus Smith's analyses are in the tradition of eighteenth-century moral sense theory and in the spirit of twentieth-century economic liberalism rather than libertarianism.

# NOTES

## Introduction

1. Joseph Schumpeter, *History of Economic Analysis*, ed. Elizabeth Boody Schumpeter (New York: Oxford University Press, 1954), p. 184. (Italics his)

2. Albert O. Hirschman, *The Passions and the Interests* (Princeton, N.J.: Princeton University Press, 1977), p. 100.

3. See Bernard Mandeville, *The Fable of the Bees or Private Vices Publick Benefits*, ed. F. B. Kaye (Oxford: Clarendon Press, 1924, reprinted Indianapolis: Liberty Classics, 1988); and Smith's commentary on Mandeville, *Theory of Moral Sentiments* (*TMS*), ed. A. L. Macfie and D. D. Raphael (Oxford: Oxford University Press, 1976, reprinted Indianapolis: Liberty Classics, 1982), VII.ii.4.6–14.

4. Adam Smith, *The Wealth of Nations* (*WN*), ed. R. H. Campbell and A. S. Skinner (Oxford: Oxford University Press, 1976, reprinted Indianapolis: Liberty Classics, 1981), I.ii.2.

5. *WN* IV.ii.9.

6. See, for example, George J. Stack, "Self-Interests and Social Value," *Journal of Value Inquiry* 18 (1984): 123–137; and Joseph Cropsey, *Polity and Economy* (Westport, Conn.: Greenwood Press, 1977).

7. *WN* IV.ix.51.

8. Horst Claus Recktenwald, "An Adam Smith Renaissance *anno* 1976? The Bicentenary Output—A Reappraisal of His Scholarship," *Journal of Economic Literature* 16 (1978): 58. (Italics his)

9. Stack, "Self-Interests and Social Value," p. 123.

10. See, for example, T. D. Campbell, *Adam Smith's Science of Morals* (London: Allen & Unwin, 1971); Knud Haakonssen, *The Science of a Legislator* (Cambridge: Cambridge University Press, 1981); Ronald Hamowy, *The Scottish Enlightenment and the Theory of Spontaneous Order* (Carbondale: Southern Illinois University Press, 1987); Istvan Hont and Michael Ignatieff, eds., *Wealth and Virtue* (Cambridge: Cambridge University Press, 1983); David Levy in a variety of articles that shall be cited in later chapters; J. R. Lindgren, *The Social Philosophy of Adam Smith*

(The Hague: Nijhoff, 1973); A. L. Macfie, *The Individual in Society: Papers on Adam Smith* (Oxford: Oxford University Press, 1967); Glenn Morrow, *The Ethical and Economic Theories of Adam Smith* (New York: A. M. Kelley, 1926; 1969); D. D. Raphael, *Adam Smith* (Oxford: Oxford University Press, 1985); John Ridpath, "Adam Smith and the Founding of Capitalism," audiotape from Conceptual Conferences, 1988; Amartya Sen, "Rational Fools," *Philosophy and Public Affairs* 6 (1977): 317–344, and *On Ethics and Economics* (Oxford: Blackwell Publisher), 1987; Andrew S. Skinner, *A System of Social Science: Papers Relating to Adam Smith* (Oxford: Clarendon Press, 1979); Richard F. Teichgraeber, III, *"Free Trade" and Moral Philosophy* (Durham, N.C.: Duke University Press, 1986); and Donald Winch, *Adam Smith's Politics* (Cambridge: Cambridge University Press, 1978). Throughout the essay I shall refer to these authors and at times take issue with some of their conclusions.

11. See Russell Nieli, "Spheres of Intimacy and the Adam Smith Problem," *Journal of the History of Ideas* 47 (1986): 611–624, for a history of the "Adam Smith Problem." It is Stack who argues that this is still the prevailing view. For other versions of this interpretation of the *WN*, see, for example, Barry Schwartz, *The Battle for Human Nature* (New York: Norton, 1986), esp. pp. 59–61, 78; Fred Hirsch, *Social Limits to Growth* (Cambridge, Mass.: Harvard University Press, 1976), p. 119; and Garrett Hardin, "The Tragedy of the Commons," *Science* 162 (1968): 1244.

12. Robert H. Frank, *Passions Within Reason* (New York: Norton, 1988), p. 21.

13. Amitai Etzioni, *The Moral Dimension* (New York: Free Press, 1988), p. 1. (Italics his)

14. Ibid., p. 22.

15. George Stigler, "Economics or Ethics?" in S. McMurrin, ed., *Tanner Lectures on Human Values* (Salt Lake City: University of Utah Press, 1981), vol. 2, p. 188.

16. Christopher Morris, "The Relation Between Self-Interest and Justice in Contractarian Ethics," in E. P. Paul, F. D. Miller, Jr., Jeffrey Paul, and John Ahrens, eds., *The New Social Contract: Essays on Gauthier* (New York: Blackwell Publisher, 1988), p. 123; and David Gauthier, *Morals by Agreement* (Oxford: Clarendon Press, 1986), pp. 86, 100, 328–329.

17. Barry Hindess, *Choice, Rationality, and Social Theory* (London: Unwin Hyman, 1988), p. 29.

18. See Jon Elster, "Introduction," in Jon Elster, ed., *Rational Choice* (Oxford: Blackwell Publisher, 1986), pp. 1–22.

19. Gauthier, *Morals by Agreement*, p. 13.

20. Ibid., p. 84. (Italics his)

21. James Buchanan, "The Gauthier Enterprise," in Paul et al., eds., *The New Social Contract*, p. 89. This may also be the view of Frank Knight. See Jules Coleman, "Competition and Cooperation" *Ethics* 97 (1987): 76–90.

22. H. T. Buckle, *History of Civilization in England* (London, 1861), vol. 2, p. 437, reprinted in "Introduction" to *TMS*, by D. D. Raphael and A. L. Macfie, p. 21.

23. See, for example, Bruno Hildebrand, *Die Nationalokonomie der Gegenwart und Zukunft* (Frankfurt, 1848); Carl G. A. Knies, *Die Politisch Oekonomie vom Standpunkte der geschichtlichen Methode* (Braunschweig, 1853); and Witold von Skarzynski, *Adam Smith als Moralphilosoph und Schoepfer der Nationaloekonomie* (Berlin, 1878), cited in "Introduction," *TMS*, pp. 21–25.

24. See especially Macfie, *The Individual in Society*.

25. For this sort of defense, see, for example, Morrow, "Introduction," *The Ethical and Economic Theories of Adam Smith*; and D. D. Raphael and A. L. Macfie, "Introduction" to their edition of the *TMS*, pp. 20–25.

26. See Jacob Viner, "Adam Smith and Laissez Faire," in J. M. Clark et al., eds., *Adam Smith, 1776–1926* (New York: A. M. Kelley, 1928, 1966), pp. 116–120. See also Jacob Viner, *The Role of Providence in the Social Order* (Philadelphia: American Philosophical Society, 1972), esp. pp. 77–86.

27. Viner, "Adam Smith," p. 134.

28. Ibid., pp. 134–137; and Viner, *The Role of Providence*, p. 82.

29. See, for example, Max Lerner's "Introduction" to Adam Smith, *The Wealth of Nations*, ed. Edwin Cannan (New York: Modern Library, 1937), p. ix.

30. George Stigler, "Smith's Travels on the Ship of State," *History of Political Economy* 3 (1971): 265. The actual quotation is "The *Wealth of Nations* is a stupendous palace erected upon the granite of self-interest."

31. Hirschman, *The Passions of the Interests*, pp. 108–111.

32. Morrow, *Ethical and Economic Theories*, p. 9.

33. *TMS* II.ii.3.4.

34. *WN* V.iii.6.

35. Stack, "Self Interests and Social Value."

36. Macfie, *The Individual in Society*, p. 101.

37. For a good summary of these problems, see Marjorie Ann Clay, "'Private Vices, Public Benefits': Adam Smith's Concept of Self-Interest," in William R. Morrow and Robert E. Stebbins, eds., *Adam Smith and the Wealth of Nations, 1776–1976* (Proceedings of the Bicentennial Conference) (Richmond: Eastern Kentucky University, 1976), pp. 42–61.

38. *TMS* II.ii.2.1.

39. Ibid.

40. P. L. Danner, "Sympathy and Exchangeable Value: Keys to Adam Smith's Social Philosophy," *Review of Social Economy* 34 (1976): 317–331.

41. Robert Boynton Lamb, "Adam Smith's System: Sympathy Not Self-Interest," *Journal of the History of Ideas*, 35 (1974): 682.

42. Macfie, *The Individual in Society*, p. 104.

43. Lamb, "Adam Smith's System," p. 681.

44. Morrow, *Ethical and Economic Theories*, p. 76.

45. Warren Samuels, "The Political Economy of Adam Smith," *Ethics* 87 (1977): 201. See also Nathan Rosenberg, "Some Institutional Aspects of the *Wealth of Nations*," *Journal of Political Economy* 68 (1960): 55–570.

46. Samuels, "Political Economy," p. 198.

47. *WN* I.ii.4.

48. *WN* IV.viii.30.

49. *WN* I.i.1.

50. *WN* V.i.f.50.

51. See, for example, Ronald Meek, *Economics and Ideology and Other Essays* (London: Chapman & Hall, 1967); and Robert Heilbroner, "The Paradox of Progress: Decline and Decay in *The Wealth of Nations*," in Andrew S. Skinner and Thomas Wilson, eds., *Essays on Adam Smith* (Oxford: Clarendon Press, 1975), pp. 524–539.

52. *WN* I.x.c.41.

53. *WN* IV.ii.3.

54. *WN* IV.v.b.16.

55. *WN* II.ii.36.

56. Heilbroner, "The Paradox of Progress," p. 524.

## Chapter 1

1. Joseph Cropsey, *Polity and Economy* (Westport, Conn.: Greenwood Press, 1977), p. 29.

2. Gregory S. Kavka, *Hobbesian Moral and Political Theory* (Princeton, N.J.: Princeton University Press, 1986), p. 39.

3. Ibid., pp. 64–81.

4. Bernard Gert, "Hobbes' Account of Reason and the Passions," in Martin Bertman and Michael LeMalherbe, eds., *Thomas Hobbes* (Paris: Libraire Philosophique J. Vrin, 1989), pp. 91–92.

5. See, for example, *TMS* I.i.2.1–2 and VII.iii.i.

6. *TMS* I.i.1.1.

7. *TMS* II.ii.2.1.

8. *TMS* I.ii.3, 4, 5.

9. *TMS* III.2.6.
10. *TMS* III.2.7.
11. *TMS* VII.iii.i.1.
12. Ralph Ansbach, "The Implications of the *Theory of Moral Sentiments* for Adam Smith's Economic Thought," *History of Political Economy* 4 (1972): 203.
13. See, for example, *TMS* III.3.3 and VII.ii.3.16. See also II.ii.2.2., "the arrogance of self-love."
14. Joseph Butler, *Fifteen Sermons Preached at Rolls Chapel* (1726, reprinted London: Thomas Tegg & Son, 1835), especially Sermons I and XI.
15. Albert O. Hirschman, *The Passions and the Interests* (Princeton, N.J.: Princeton University Press, 1977), pp. 100–113.
16. *TMS* II.i.5.8–10 (note to II.i.5.6).
17. *TMS* I.ii.3.8.
18. See Louis Schneider, "Adam Smith on Human Nature and Social Circumstance," in Gerald P. O'Driscoll, Jr., ed., *Adam Smith and Modern Political Economy* (Ames: Iowa State University Press, 1979), pp. 44–69.
19. *TMS* VII.ii.3.16.
20. Ibid.
21. *TMS* VII.iii.1.4.
22. Milton Friedman, "Adam Smith's Relevance for 1976," University of Chicago Graduate School of Business Occasional Papers 50, pp. 16–17.
23. Robert Boynton Lamb, "Adam Smith's System: Sympathy Not Self-Interest," *Journal of the History of Ideas* 35 (1974): 682.
24. For example, Cropsey says (that Smith argues): "Every human being has the power to feel the passions of those other beings who come under his observation. The man who observes joy of another will himself experience joy. . . . The very words right and wrong have and can have no other meaning than what by our emotions we sympathize with" (pp. 12–13).
25. *TMS* I.i.1.2.
26. *TMS* I.i.1.5. See also T. D. Campbell, *Adam Smith's Science of Morals* (London: Allen & Unwin, 1971), chap. 4.
27. Cropsey tries to defend that identification (*Polity and Economy*, pp. 12–13).
28. Campbell, *Adam Smith's Science of Morals*, p. 172.
29. *TMS* VII.iii.i.4.
30. *TMS* I.i.1.10.
31. *TMS* I.i.2.title.
32. *TMS* I.iii.1.9, note.
33. *TMS* VII.iii.3.16.
34. *TMS* III.4.8.
35. See, for example, David Levy, "Adam Smith's Natural Law and Social Contract," *Journal of the History of Ideas* 39 (1978): 665–674.

36. See Adam Smith, *Lectures on Jurisprudence*, ed. R. L. Meek, D. D. Raphael, and P. G. Stein (Oxford: Oxford University Press, 1978, reprinted Indianapolis: Liberty Classics, 1982), hereafter referred to as *LJ*. Citations in the text are from the report dated 1762-3 (*LJ(A)*) or the report dated 1763-4 (*LJ(B)*). See, for example, *LJ(A)* 120–127 and *LJ(B)* 94–95 on this point.

37. *TMS* III.2.2.

38. Gilbert Harman, *Moral Agent and Impartial Spectator: The Lindsey Lecture* (Lawrence: University of Kansas Press, 1986), pp. 8–14.

39. *TMS* III.1.2.

40. *TMS* III.3.4.

41. Harman, *Moral Agent and Impartial Spectator*.

42. For a good discussion of the impartial spectator, see Campbell, *Adam Smith's Science of Morals*, pp. 127–135.

43. Mark Waymack, "Moral Philosophy and Newtonianism in the Scottish Enlightenment," Ph.D. diss., Johns Hopkins University, 1987.

44. See, however, Charles L. Griswald, Jr., "Adam Smith on Virtue and Self-Interest," *Journal of Philosophy* 86 (1989): 681.

45. *TMS* VII.ii.1.50.

46. *TMS* I.1.5.7.

47. *TMS* VI.ii.1.1.

48. *TMS* I.i.5.5.

49. *TMS* VII.ii.3.

50. *TMS* VI.ii.3.6.

51. *TMS* II.ii.3.4.

52. *TMS* II.ii.1.9.

53. *TMS* II.ii.2.1.

54. *TMS* II.ii.2.2. Also see Campbell, *Adam Smith's Science of Morals*, pp. 186–190.

55. *TMS* III.6.10.

56. *TMS* II.ii.3.3.

57. Glenn Morrow, *The Ethical and Economic Theories of Adam Smith* (New York: A. M. Kelley, 1969), p. 46.

58. Ibid., p. 54.

59. *TMS* VI.iii.11.

60. *TMS* VI.iii.1.

61. *TMS* VI.i.15.

62. *TMS* VI.conclusion 6.

63. See Campbell, *Adam Smith's Science of Morals*, p. 168; and *TMS* VI.iii.12ff.

64. *TMS* VII.ii.3.

65. *TMS* IV.2.3.

66. *TMS* IV.1.1.
67. *TMS* II.iii.introduction 3.
68. *TMS* II.iii.3.2.
69. *TMS* IV.2.4.
70. J. Ralph Lindgren, "Adam Smith's Theory of Inquiry," *Journal of Political Economy* 77 (1969): p. 897. Lindgren himself does not subscribe to this interpretation of Smith's method.
71. For this interpretation, see, for example, Cropsey, *Polity and Economy*; and Waymack, "Moral Philosophy."
72. *TMS* V.1.2.
73. Lindgren, "Adam Smith's Theory of Inquiry," p. 913.
74. Cropsey, *Polity and Economy*, pp. 2–3.
75. *TMS* II.i.5.10.
76. *TMS* II.ii.2.1; see also VI.ii.1.1.
77. *TMS* II.ii.3.1.
78. *TMS* VI.ii.introduction 2.
79. For an "organic" interpretation of Smith, see, for example, John Ridpath, "Adam Smith and the Founding of Capitalism."
80. See, for example, Waymack, "Moral Philosophy"; and A. L. Macfie, *The Individual in Society: Papers on Adam Smith* (Oxford: Oxford University Press, 1967).
81. *TMS* III.1.3.
82. Campbell, *Adam Smith's Science of Morals*, p. 144.
83. See ibid., pp. 139–145.
84. See Jacob Viner, "Adam Smith and Laissez Faire," in J. M. Clark et al., eds., *Adam Smith, 1776–1926* (New York: A. M. Kelley, 1966), p. 117.
85. *TMS* III.5.12.
86. See David K. Hart, "Adam Smith in the Twentieth Century: Is the Vision Intact?" *Exchange*, Winter 1985, pp. 29ff, and *A Moral Theory of Capitalism*, forthcoming.

## Chapter 2

1. *LJ(A)* i.9.
2. See, for example, Knud Haakonssen, *The Science of a Legislator* (Cambridge: Cambridge University Press, 1981); Istvan Hont and Michael Ignatieff, eds., *Wealth and Virtue* (Cambridge: Cambridge University Press, 1983); Richard F. Teichgraeber, III, *"Free Trade" and Moral Philosophy* (Durham, N.C.: Duke University Press, 1986); and Donald Winch, *Adam Smith's Politics* (Cambridge: Cambridge University Press, 1978).

3. *WN* V.i.b.25.

4. *WN* IV.ix.17.

5. *WN* V.i.b.2.

6. *WN* V.i.b.12.

7. Istvan Hont and Michael Ignatieff, "Needs and Justice in the *Wealth of Nations*: An Introductory Essay," in Hont and Ignatieff, eds., *Wealth and Virtue*, p. 4.

8. See, for example, Haakonssen, *The Science of a Legislator*, p. 99; and Winch, *Adam Smith's Politics*. See also *TMS* II.ii.1–2.

9. This connection was pointed out to me by David Levy. See also David Hume, *A Treatise of Human Nature* (*THN*), ed. L. A. Selby-Bigge (Oxford: Clarendon Press, 1888, 1960), bk. III, pt. II, pp. 477–573.

10. *LJ(B)* 11. See also *LJ(A)* i.24–25.

11. *LJ(A)* i.12, note 3.

12. *LJ(A)* i.9.

13. *LJ(A)* i.14.

14. *LJ(A)* i.14–15.

15. Francis Hutcheson, *A Short Introduction to Moral Philosophy*, II.4.2, reprinted in *LJ(A)* i.12 note 3.

16. *LJ(A)* i.13.

17. *LJ(A)* 1–9 and *LJ(B)* 8.

18. *LJ(A)* i.13.

19. *WN* IV.ix.51.

20. Adolph Lowe, "Adam Smith's System of Equilibrium Growth," in Andrew S. Skinner and Thomas Wilson, eds., *Essays on Adam Smith* (Oxford: Clarendon Press, 1975), pp. 424–425.

21. See Joseph Cropsey, "The Invisible Hand and Moral and Political Considerations," in Gerald P. O'Driscoll, Jr., ed., *Adam Smith and Modern Political Economy* (Ames: Iowa State University Press, 1979), pp.165–176.

22. *WN* I.x.a.1.

23. *WN* I.vii.9 and I.vii.30.

24. See Joseph Cropsey, *Polity and Economy* (Westport, Conn.: Greenwood Press, 1977), p. 72. Cropsey also repeats this observation on p. 79.

25. *WN* I.xi.p.8–10. See also Winch, *Adam Smith's Politics*; and Haakonssen, *The Science of a Legislator*.

26. *LJ(A)* i.25.

27. See Haakonssen, *The Science of a Legislator*, chap. 6, for an excellent discussion of Smith's notion of property rights and their perfect duties. Haakonssen does not explain how Smith justifies his claim that property rights are both adventitious and perfect rights.

28. *WN* I.xi.c.27. See also I.x.c.12.

29. *WN* I.x.c.12.
30. See Winch, *Adam Smith's Politics*, pp. 58–59; and Haakonssen, *The Science of a Legislator*, pp. 106–107.
31. See John Locke, *Two Treatises of Government*, ed. Peter Laslett (Cambridge: Cambridge University Press, 1983), *Second Treatise*, chap. 5, pp. 303–320.
32. Robert Boynton Lamb, *Property, Markets, and the State in Adam Smith's System* (New York: Garland Press, 1987), chap. 3.
33. Ibid.
34. John Tully, *A Discourse on Property: John Locke and His Adversaries* (Cambridge: Cambridge University Press, 1980). The brunt of Tully's attack is C. B. Macpherson who, according to Tully, attributes to Locke a labor theory of property and, in civil society, the right to the acquisition of unlimited property. Both of these attributions are erroneous, according to Tully, and rather than being a "precapitalist," Locke is firmly in the natural law tradition. See also C. B. Macpherson, *The Political Theory of Possessive Individualism* (Oxford: Oxford University Press, 1972).
35. See C. B. Macpherson, *Democratic Theory* (Oxford: Clarendon Press, 1975), pp. 123–135, for this distinction.
36. Tully, *A Discourse on Property*, p. 131.
37. Locke, *Second Treatise*, chap. 5, pp. 25–36.
38. Tully, *A Discourse on Property*, p. 122.
39. See ibid., p. 149, pp. 157–176.
40. Hume, *THN*, pp. 495–496.
41. *WN* V.i.b.2.
42. *LJ(A)* iv.19.
43. *LJ(A)* i.17.25.
44. *LJ(A)* i.35–36, 43, 77, 90; and *LJ(B)* 150, 154.
45. *LJ(A)* i.27–2.13; and *LJ(B)* 149–171.
46. *LJ(A)* ii.14–56; and *LJ(B)* 172–181.
47. A. L. Macfie, *The Individual in Society: Papers on Adam Smith* (Oxford: Oxford University Press, 1967), pp. 83–87. See also *TMS* III.4.5–10 and VII.iii.2.6.
48. See Teichgraeber, "*Free Trade*," p. 154; and Haakonssen, *The Science of a Legislator*, chap. 8.
49. *LJ(B)* 19.
50. *LJ(B)* 20.
51. *LJ(A)* iv.41–74; and *LJ(B)* 20–22.
52. *LJ(B)* 22–26.
53. *WN* III.iv.14.
54. See *WN* III.iv; and Teichgraeber, "*Free Trade*," pp. 151–152.

55. See Nathan Rosenberg, "Adam Smith, Consumer Tastes, and Economic Growth," *Journal of Political Economy* 76 (1968): 368–371, for a discussion of economic growth that brings on the age of commerce.

56. *WN* V.i.b.2.

57. See Thomas J. Lewis, "Adam Smith: The Labor Market as the Basis for Natural Right," *Journal of Economic Issues* 11 (1977): 21–50.

58. See, for example, Ronald L. Meek, "The Rise and Fall of the Concept of the Economic Machine," chap. 2 of *Economics and Ideology and Other Essays* (Leicester: Leicester University Press, 1965); Robert Heilbroner, "The Paradox of Progress: Decline and Decay in *The Wealth of Nations*," in Skinner and Wilson, eds., *Essays on Adam Smith*; and Robert Heilbroner, "The Man and His Times," in Robert Heilbroner, ed., *The Essential Adam Smith* (New York: Oxford University Press, 1986), pp. 1–11.

59. *LJ(B)* 20.

60. *LJ(B)* 11.

61. See Andrew Skinner, "A Scottish Contribution to Marxist Sociology?" in Ian Bradley and Michael Howard, eds., *Classical and Marxian Political Economy* (essays in honor of Ronald Meek) (New York: St. Martin's Press, 1981), pp. 78–114, for a critique of the view that Smith is a historical materialist.

62. See Jacob Viner, "Adam Smith and Laissez Faire," in J. M. Clark et al., eds., *Adam Smith, 1776–1926* (New York: A. M. Kelley, 1966), pp. 134–138, for a list of these failures.

63. For example, the *Index of Legal Periodicals* uses this term to index employment relationships until the end of 1980.

64. *WN* I.x.c.12.

65. See Christopher Morris, "The Relation Between Self-Interest and Justice in Contractarian Ethics," in E. P. Paul, F. D. Miller, Jr., Jeffrey Paul, and John Ahrens, eds., *The New Social Contract: Essays on Gauthier* (New York: Blackwell Publisher, 1988), pp. 118–130, for a discussion of the difficulties with a social contract analysis.

66. *LJ(A)* v.129. See also iv.19, v.115–116, 134–135; *LJ(B)* 15, 18. 91–99.

67. *LJ(A)* v.128.

68. *TMS* I.iii.2.1; and *LJ(B)* 12–13.

69. *LJ(A)* v.120–127; *LJ(B)* 94–95; and *WN* I.xi.pp.8–9. See also Haakonssen, *The Science of a Legislator*, esp. chap. 5; and Winch, *Adam Smith's Politics*, chap. 3, for good discussions of Smith's criticisms of the original contract theorists.

70. *LJ(A)* V.127.

71. *LJ(B)* 94.

72. *LJ(B)* 95.

73. Lamb, *Property, Markets*, chap. 8. See also Denis Collins, "Adam Smith's Social Contract," *Business and Professional Ethics Journal* 7 (1988): 119–146.

74. See H. J. Bitterman, "Adam Smith's Empiricism and the Law of Nature," pt. 1, *Journal of Political Economy* 48 (1940): 487–520.

75. *LJ(A)* i.1.

76. *LJ(B)* 5.

77. *LJ(B)* 5.

78. *LJ(A)* i.1.

79. *TMS* II.ii.3.4.

80. *TMS* II.ii.2.1.

81. *WN* IV.ix.16. See also IV.ix.17 and I.x.c.27.

82. *TMS* VII.iv.36.

83. *LJ(A)* v.142.

84. *LJ(A)* i.14–16, v. 142.

85. *WN* IV.viii.30.

86. *WN* I.x.a.1; I.x.c.12, 42–59; II.ii.94; IV.v.a.23; IV.v.b.16; and IV.vii.c.44, 59–62.

87. *WN* V.iii.7.

88. See *WN* I.viii.12, I.ix.24, and I.x.c.25,27.

89. *WN* IV.ix.51.

90. *WN* II.ii.94.

91. Hume, *THN*, pp. 475, 483–484, 579–580.

92. *TMS* II.ii.3.7–12. See also Haakonssen, *The Science of a Legislator*, pp. 87–89.

93. Teichgraeber, "*Free Trade*," pp. 154–157.

94. *TMS* III.3.11.

95. *TMS* VI.ii.2.16.

96. See Ronald Hamowy, *The Scottish Enlightenment and the Theory of Spontaneous Order* (Carbondale: Southern Illinois University Press, 1987), p. 15; and Teichgraeber, "*Free Trade*," pp. 139–170.

97. *WN* IV.ix.51.

98. George J. Stack, "Self-Interests and Social Value," *Journal of Value Inquiry* 18 (1984): 124.

99. *WN* V.i.b.25.

100. Leonard Billet, "The Just Economy: The Moral Basis for the *Wealth of Nations*," *Review of Social Economy* 34 (1976): 304–306.

101. *WN* I.viii.36, I.x.a.1, I.x.b.40, I.x.c.41–42, IV.viii.30; and IV.ix.3.

102. *WN* I.ii.4.

103. *WN* I.viii.36.

104. *WN* I.viii.13–14.

105. *WN* I.vi.8.

106. *WN* I.x.a.1.

107. David Levy, "David Hume's Invisible Hand in *The Wealth of Nations*: The Public Choice of Moral Information," *Hume Studies* II (1985): 110.

108. *LJ(B)* 326.

109. See Jeremy Shearmur, "Adam Smith's Second Thoughts: Economic Liberalism and Its Unintended Consequences," *The Adam Smith Club Kirkaldy Paper No. 1.*

110. Haakonssen, *The Science of a Legislator*, p. 105.

## Chapter 3

1. George Stigler, "Smith's Travels on the Ship of State," *History of Political Economy* 3 (1971): 265.

2. See, for example, Horst Claus, Recktenwald, "An Adam Smith Renaissance *anno* 1976?" The Bicentenary Output—A Reappraisal of His Scholarship," *Journal of Economic Literature* 16 (1978): 58.

3. See, for example, the writers listed in the Introduction, note 10.

4. A. L. Macfie, *The Individual in Society: Papers on Adam Smith* (Oxford: Oxford University Press, 1967), p. 101.

5. Alan S. Blinder, *Hard Heads Soft Hearts* (Reading, Mass.: Addison-Wesley, 1987), p. 27.

6. Stigler, "Smith's Travels."

7. Lionel Robbins, *Political Economy: Past and Present* (London: Macmillan, 1963), pp. 34–35.

8. *TMS* VI.ii.1.1.

9. *WN* II.iii.28. See also *WN* IV.v.b.16, 43.

10. *WN* I.ii.2.

11. *WN* I.ii.2.

12. David Gauthier, *Morals by Agreement* (Oxford: Clarendon Press, 1986), p. 87.

13. See P. H. Wicksteed, *The Common Sense of Political Economy*, ed. Lionel Robbins (London: G. Routledge & Sons, 1946), vol. 1, p. 180.

14. Russell Nieli, "Spheres of Intimacy and the Adam Smith Problem," *Journal of the History of Ideas* 47 (1986): 611–624.

15. Max Lerner, "Introduction" to Adam Smith, *The Wealth of Nations*, ed. Edwin Cannan (New York: Modern Library, 1937), p. ix.

16. Albert O. Hirschman, *The Passions and the Interests* (Princeton, N.J.: Princeton University Press, 1977), p. 100.

17. See *WN* I.ix.24, I.x.c.25–26, I.xi.p.10, II.iii.25–26, IV.vii.c.61, and V.i.b.2.

18. *WN* II.iii.25.

19. *WN* II.ii.36.

20. *WN* I.xi.c.31, II.iii.29–30, IV.ix.13, and V.i.b.2. See also *TMS* I.iii.3.

21. *WN* I.ii.2.

22. P. L. Danner, "Sympathy and Exchangeable Value: Keys to Adam Smith's Social Philosophy," *Review of Social Economy* 34 (1976): 324.

23. *WN* I.ii.1.

24. *WN* III.i.1.

25. *WN* II.v.3.

26. *WN* IV.ix.16–25.

27. *WN* I.x.c.27.

28. *WN* III.i.1.

29. *WN* I.xi.p.8–10 and IV.vii.c.87–88.

30. See Glenn Morrow, *The Ethical and Economic Theories of Adam Smith* (New York: A. M. Kelley, 1969), p. 75; and Marjorie Ann Clay, "'Private Vices, Public Benefits': Adam Smith's Concept of Self Interest," in William R. Morrow and Robert E. Stebbins, eds., *Adam Smith and the Wealth of Nations, 1776–1976* (Proceedings of the Bicentennial Conference) (Richmond: Eastern Kentucky University, 1976), pp. 49–50.

31. Clay, "'Private Vices, Public Benefits,'" pp. 49–50.

32. Danner, "Sympathy and Exchangeable Value," p. 322. (Italics his)

33. Ibid., p. 324.

34. *WN* I.v.17.

35. Robert Boynton Lamb, "Adam Smith's System: Sympathy Not Self-Interest," *Journal of the History of Ideas* 35 (1974): 682.

36. *WN* I.xi.c.27.

37. *TMS* I.iii.3.

38. Herbert W. Schneider, ed., *Adam Smith's Moral and Political Philosophy* (New York: Hafner, 1948), p. xx.

39. *WN* IV.ix.51.

40. *TMS* IV.I.10.

41. *WN* IV.ii.9.

42. Macfie, *The Individual in Society*, p. 101.

43. Lamb, "Adam Smith's System," p. 681.

44. See T. D. Campbell, *Adam Smith's Science of Morals* (London: Allen & Unwin, 1971), p. 72.

45. *TMS* VI.ii.2.17. See also *TMS* IV.I.11.

46. *WN* IV.ix.51.

47. *WN* IV.v.b.16 and IV.ix.49.

48. *WN* I.xi.p.10.
49. *WN* I.viii.13.
50. *WN* IV.ii.21 and IV.v.a.23.
51. *WN* IV.vii.c.62.
52. For example, *WN* I.x.c and II.v.7–12.
53. *WN* I.x.a.1.
54. See, for example, *WN* I.x.c.1. and V.i.b.2.
55. *WN* I.ix.16–17.
56. *WN* II.ii.94.
57. *WN* I.x.c.27.
58. David K. Hart, *A Moral Theory of Capitalism*, forthcoming.
59. *WN* I.x.c.27 and I.xi.p.
60. *WN* IV.ii.9.
61. Blinder, *Hard Heads Soft Hearts*, p. 27.
62. James Buchanan, "The Gauthier Enterprise," in E. P. Paul, F. D. Miller, Jr., Jeffrey Paul, and John Ahrens, eds., *The New Social Contract: Essays on Gauthier* (New York: Blackwell Publisher, 1988), p. 89.

## Chapter 4

1. Glenn Morrow, *The Ethical and Economic Theories of Adam Smith* (New York: A. M. Kelley, 1969), p. 76.
2. L.T. Elzie, "Self Interest and Economic Power," in William R. Morrow and Robert E. Stebbins, eds., *Adam Smith and the Wealth of Nations, 1776–1976* (Proceedings of the Bicentennial Conference) (Richmond: Eastern Kentucky University, 1976), p. 18.
3. Lionel Robbins, *The Theory of Economic Policy in English Classical Political Economy* (London: Macmillan, 1953); Lionel Robbins, *Political Economy: Past and Present* (London: Macmillan, 1963); Nathan Rosenberg, "Some Institutional Aspects of the *Wealth of Nations*," *Journal of Political Economy* 68 (1960): 557–570; Warren Samuels, "The Political Economy of Adam Smith," *Ethics* 87 (1977): 189–207; and Warren Samuels, "Adam Smith and the Economy as a System of Power," *Review of Social Economy* 31 (1973): 123–137.
4. Rosenberg, "Some Institutional Aspects," p. 559.
5. Samuels, "The Political Economy of Adam Smith," pp. 199–200.
6. Ibid., p. 202. (Italics his)
7. *TMS* II.ii.3.1.
8. Morrow, *Ethical and Economic Theories*, p. 40.
9. Ibid., p. 12.
10. Ibid., pp. 71–72.

11. Ibid., p. 42.

12. Ibid., p. 84.

13. See also Gladys Bryson, *Man and Society* (Princeton, N.J.: Princeton University Press, 1945), pp. 160–161, for confirmation of this theory. J. R. Lindgren discusses this issue in chap. 3 of *The Social Philosophy of Adam Smith* (The Hague: Nijhoff, 1973). For further discussions of individualism and holism, see Leon J. Goldstein, "The Two Theses of Methodological Individualism," *British Journal for the Philosophy of Science* 9 (1958): 1–11; Maurice Mandelbaum, "Societal Laws," *British Journal for the Philosophy of Science* 8 (1957): 211–224; and Joseph Agassi, "Methodological Individualism," *British Journal of Sociology* 26 (1975): 244–270.

14. Robbins, *Political Economy*, p. 6.

15. Ibid., pp. 34–35, citing *WN* I.i.10 and I.ii.2.

16. Robbins, *The Theory of Economic Policy*, p. 187.

17. Samuels, "The Political Economy of Adam Smith," p. 192.

18. *WN* V.i.f.7. See also *WN* I.xi.p.8, V.i.b.2, and V.i.d.5.

19. Rosenberg, "Some Institutional Aspects," p. 560.

20. Rosenberg, "Some Institutional Aspects," pp. 569–570.

21. *WN* V.i.b.20. See also V.i.f.4–5.

22. Rosenberg, "Some Institutional Aspects," p. 563.

23. *WN* V.i.f.50–53.

24. Samuels, "The Political Economy of Adam Smith," p. 198.

25. Warren J. Samuels, *The Classical Theory of Economic Policy* (Cleveland: World Publishing, 1966), esp. chap. 2.

26. Samuels, "The Political Economy of Adam Smith" and "Adam Smith and the Economy as a System of Power."

27. Samuels, "Adam Smith and the Economy as a System of Power," p. 123.

28. *WN* I.xi.p.1.

29. *WN* I.viii.11–14.

30. *WN* I.xi.p.10.

31. Samuels, "Adam Smith and the Economy as a System of Power," pp. 123–137.

32. *WN* IV.ix.51.

33. *TMS* II.ii.1.5, III.6.8–11, VII.iv.1, and *LJ(A)* i.14–18.

34. *LJ(A)* i.14–15.

35. Rosenberg, "Some Institutional Aspects," p. 560.

36. See *WN* V.i.d for a discussion of individual payments for the use of public works such as highways and bridges.

37. Samuels, "The Political Economy of Adam Smith," pp. 201–202. (Italics his) See also Glenn Morrow, "Adam Smith: Moralist and Philoso-

pher," in J. M. Clark et al., eds., *Adam Smith, 1776–1926* (New York: A. M. Kelley, 1966), p. 172.

38. Goldstein, "The Two Theses of Methodological Individualism," pp. 1–11; and Anthony Quinton, "Social Objects," *Proceedings of the Aristotelian Society* 76 (1976): pp. 1–27. See also J. W. N. Watkins, "Historical Explanation in the Social Sciences," *British Journal of the Philosophy of Science* 7 (1957): 104–117.

39. Maurice Mandelbaum, "Societal Facts," *British Journal of Sociology* 6 (1955): 307.

40. Agassi, "Methodological Individualism," pp. 264–268.

41. See Goldstein, "The Two Theses of Methodological Individualism," for a good explanation of these four views. See also Leon Goldstein, "The Inadequacy of the Principle of Methodological Individualism," *Journal of Philosophy* 53 (1956): 801–813; and J. W. N. Watkins, "The Alleged Inadequacy of Methodological Individualism," *Journal of Philosophy* 64 (1957): 390–395.

42. Warren Samuels, from a review of an early version of this chapter.

43. *WN* V.i.b.14–15, d.5, f, g.

44. *WN* I.ix.15–16. See also *WN* V.i.e.

45. *WN* IV.viii.30.

46. *WN* V.i.e.2–5, 18. See also Joseph J. Spengler, "Adam Smith's Theory of Economic Growth, Part I," *Southern Economic Journal* 21 (1959): 411–415.

47. See *TMS* VI.ii.2.17.

48. *WN* II.v.7. and II.x.c.28–32.

49. *WN* II.ii.36.

50. *WN* II.ii.94.

51. *WN* V.i.f.46–50 and V.i.g.15–16. See also Frank Petrella, "Individual, Group or Government? Smith, Mill, and Sidgwick," *History of Political Economy* 2 (1970): 170–176.

52. *TMS* III.1.3.

53. *TMS* II.ii.3.1.

54. *TMS* II.ii.3.10.

55. *TMS* VI.ii.2.17.

56. *TMS* VI.ii.2.8.

57. See, for example, *TMS* II.ii.3.4..

58. *TMS* II.ii.2.1. See also II.i.5.10 and VI.i.1.

59. *WN* IV.ii.9.

60. Samuels, "Adam Smith and the Economy as a System of Power," p. 125.

61. *WN* I.x.c.27, I.xi.p, IV.ii.9, and V.i.d.

62. *WN* IV.viii.30, IV.ix.3, and V.i.b.25.

63. *WN* V.i.b.2.

64. *WN* V.i.b.12.

65. For example, *WN* IV.ii.3–4, IV.v.a.23, and IV.v.b.7, 16.

66. *WN* I.x.c.1–2, 27–33, II.iii.29, and II.v.7–8. See also Irvin Sobel, "Adam Smith: What Kind of Institutionalist Was He?" *Journal of Economic Issues* 13 (1979): 347–368.

## Chapter 5

1. *WN* Introduction, 1.

2. Joseph Schumpeter, *History of Economic Analysis*, ed. Elizabeth Boody Schumpeter (New York: Oxford University Press, 1954), p. 187.

3. Whether Smith was dependent on Adam Ferguson's analysis of the division of labor is a matter of debate among Smith scholars. Marx attributes Smith's analysis to Ferguson, but according to Smith's biographer, John Rae, Smith accused Ferguson of borrowing his ideas. See Ronald Hamowy, "Adam Smith, Adam Ferguson, and the Division of Labour," *Economica* 35 (1968): pp. 249–259; and Jacob Viner, "Guide to John Rae's *Life of Adam Smith*," in John Rae, *Life of Adam Smith* (New York: A. M. Kelley, 1895, 1965), pp. 35–36.

4. *WN* I.i.1.

5. *WN* V.i.f.50.

6. For example, Robert Heilbroner, "The Paradox of Progress: Decline and Decay in the *Wealth of Nations*," in Andrew S. Skinner and Thomas Wilson, eds., *Essays on Adam Smith* (Oxford: Clarendon Press, 1975), pp. 524–539; and Ronald Meek, *Studies in the Labour Theory of Value* (London: Lawrence & Wishart, 1956, 1973).

7. Leonard Krieger, *Kings and Philosophers, 1689–1789* (New York: Norton, 1970), p. 199.

8. *WN* I.x.c.12.

9. John Tully, *A Discourse on Property: John Locke and His Adversaries* (Cambridge: University Press, 1980), pp. 136–143.

10. *WN* III.iv.1–18. See also Thomas J. Lewis, "Adam Smith: The Labor Market as the Basis of Natural Right," *Journal of Economic Issues* 11 (1977): 24–39.

11. *WN* I.vi.1–4.

12. *LJ(A)* i.10–18.

13. *WN* III.iv.10–12.

14. *WN* III.ii.12 and IV.ix.47.

15. *WN* III.iv. See also Lewis, "Adam Smith: The Labor Market," pp. 21–50.

16. *WN* I.i.8.

17. *WN* II.Introduction and II.i. See P. D. Groenewegen, "Adam Smith and the Division of Labour: A Bicentenary Estimate," *Australian Economic Papers* 16 (1977): 161–165, for a nice summary of Smith's theory of the division of labor.

18. *WN* IV.ii.9.

19. *WN* I.v.1.

20. See Meek, *Studies in the Labour Theory of Value*, esp. pp. 44–62.

21. *WN* I.vi.9.

22. *WN* I.v.9. See also *LJ(A)* vi.36: "We are not to judge whether labour be cheap or dear by the moneyd price of it, but by the quantity of the necessaries of life which may be got by the fruits of it."

23. *WN* I.v.7.

24. *WN* I.v.17.

25. See Paul Douglas, "Smith's Theory of Value and Distribution," in J. M. Clark et al., eds., *Adam Smith, 1776–1926* (New York: A. M. Kelley, 1928, 1966), pp. 88–103.

26. *WN* V.i.f.50. Smith may have discovered some of his worries about the negative consequences of the division of labor in Rousseau, although his analysis of these consequences is obviously quite different. See Hamowy, "Adam Smith, Adam Ferguson, and the Division of Labour," pp. 258–259.

27. See, for example, E. G. West, "Adam Smith's Two Views on the Division of Labour," *Economica* 31 (1964): 23–32; and Heilbroner, "The Paradox of Progress," pp. 524–539.

28. See Nathan Rosenberg, "Adam Smith, Consumer Tastes, and Economic Growth," *Journal of Political Economy* 76 (1968): 361–374.

29. See Istvan Hont and Michael Ignatieff, "Needs and Justice," in Istvan Hont and Michael Ignatieff, eds., *Wealth and Virtue* (Cambridge: Cambridge University Press, 1983), pp. 1–44.

30. See Heilbroner, "The Paradox of Progress" and Meek, *Studies in the Labour Theory of Value*.

31. See especially Karl Marx, *Economic and Philosophical Manuscripts*, in T. B. Bottomore, trans. and ed., *Early Writings* (New York: McGraw-Hill, 1963), pp. 61–220; and Karl Marx, *Grundrisse*, trans. and ed. David McLellan (New York: Harper & Row, 1971).

32. *WN* V.i.f.46–57.

33. See E. G. West, "The Political Economy of Alienation: Karl Marx and Adam Smith," *Oxford Economic Papers* 21 (1969): 1–23. West claims that Smith recognizes the problem of self-estrangement in the specialization of labor, but he does not conclude that Smith solves that problem.

34. *WN* I.xi.p.9.

35. *WN* V.i.f.61.
36. *WN* V.i.f.51.
37. Marx, *Economic and Philosophical Manuscripts*, third manuscript, pp. 147–151.
38. Smith even goes so far as to say: "It is the great multiplication of the production of all the different arts, in consequence of the division of labour, which occasions, in a well- governed society, that universal opulence which extends itself to the lowest ranks of the people" (*WN* I.i.10).
39. See E. G. West, "Adam Smith and Alienation: Wealth Increases, Men Decay?" in Andrew S. Skinner and Thomas Wilson, eds., *Essays on Adam Smith* (Oxford: Clarendon Press, 1975), pp. 540–552.
40. See, for example, David Ellerman, "On the Labor Theory of Property," *Philosophical Forum* 16 (1985): 293–326, for this view.
41. *TMS* VI.ii.2.17.
42. *WN* I.x.c.41.
43. *WN* I.x.c.59.
44. *WN* I.x.c.61.
45. *WN* I.x.a.1.
46. *WN* I.x.b.40–52.
47. *WN* I.x.b.1.
48. *WN* I.viii.44.
49. *WN* I.viii.45–48.
50. *WN* I.x.c.61.
51. *WN* I.x.c.61.
52. *WN* II.ii.94.
53. *WN* II.iv.14.
54. *WN* I.x.c.61.
55. Jacob Viner, *The Role of Providence in the Social Order* (Philadelphia: American Philosophical Society, 1972), p. 82.
56. *WN* I.viii.36.
57. *WN* I.ix.23.
58. *WN* I.xi.p.
59. *WN* I.ii.5 and V.i.b.1–11.

## Chapter 6

1. Ralph Lindgren depicts many modern interpretations of the *WN* this way, although he argues that this picture of the *WN* is false. Although he is partially correct, he may have exaggerated the falsity of this claim. See J. R. Lindgren, *The Social Philosophy of Adam Smith* (The Hague: Nijhoff, 1973), p. 84.

2. *WN* IV.Introduction.
3. *WN* IV.vii.c.44.
4. See Richard F. Teichgraeber, III, *"Free Trade" and Moral Philosophy* (Durham, N.C.: Duke University Press, 1986), p. 156; and Jacob Viner, "Adam Smith and Laissez-Faire," in J. M. Clark et al., eds., *Adam Smith, 1976–1926* (New York: A. M. Kelley, 1928, 1966), pp. 133–134.
5. Joseph Schumpeter, *History of Economic Analysis*, ed. Elizabeth Boody Schumpeter (New York: Oxford University Press, 1954), p. 147, n. 5.
6. *WN* IV.i.1.
7. *WN* IV.viii.1.
8. See Jacob Viner, "Mercantilist Thought," in the *International Encyclopedia of the Social Sciences, Volume 4*, ed. David L. Sills (New York: Macmillan, 1936, 1968), pp. 435–443. My knowledge of mercantilism also benefited from many discussions with my colleague Tom Donaldson.
9. *WN* IV.i.1.
10. *WN* IV.i.31.
11. *WN* IV.ii.11.
12. *WN* IV.v.a.
13. *WN* IV.viii.30. See also *WN* IV.vii.c.60–62.
14. *WN* IV.viii.54.
15. D. C. Coleman, "Adam Smith, Businessmen, and the Mercantile System in England," *History of European Ideas* 9 (1988): 165–170. See also A. W. Coats, "Adam Smith and the Mercantile System," in Andrew S. Skinner and Thomas Wilson, eds., *Essays on Adam Smith* (Oxford: Clarendon Press, 1975), pp. 219–236.
16. See Schumpeter, *History of Economic Analysis*, pp. 361–362; and Coleman, "Adam Smith, Businessmen, and the Mercantile System," pp. 161–170.
17. See R. L. Meek, *The Economics of Psysiocracy* (London: Allen & Unwin, 1962), esp. chap. 1.
18. *WN* IV.ix.29.
19. *WN* I.iv.13.
20. *WN* IV.vii.c.44,88.
21. Viner, "Adam Smith and Laissez Faire," pp. 133–134.
22. *WN* II.ii.106, II.v.7, and IV.vii.c.87–89.
23. *WN* II.ii.106.
24. *WN* IV.v.b.16.
25. *WN* IV.v.b.43.
26. *LJ(B)* 12.
27. *WN* I.xi.p.9.
28. *WN* I.xi.p.8.
29. *WN* I.xi.p.10.

30. *WN* V.i.e.18.

31. L. T. Elzie, "Self Interest and Economic Power," in William R. Morrow and Robert E. Stebbins, eds., *Adam Smith and the Wealth of Nations, 1776–1976* (Proceedings of the Bicentennial Conference)(Richmond: Eastern Kentucky University, 1976), p. 18.

32. *WN* IV.ix.51.

33. *WN* V.i.c.1.

34. *WN* II.ii.94.

35. *WN* IV.vii.c.60–64.

36. *WN* IV.ix.16. See also IV.ix.17 and I.x.c.27.

37. Warren S. Gramm, "The Selective Interpretation of Adam Smith," *Journal of Economic Issues* 14 (1980): 128.

38. *WN* V.i.e.18.

39. See also Warren S. Gramm, "The Eighteenth Century Adam Smith," Lecture at Ripon College, March 1988.

40. Gunnar Myrdal, *The Political Element in the Development of Economic Theory* (London: Routledge & Kegan Paul, 1953), p. 107, requoted in Nathan Rosenberg, "Some Institutional Aspects of the *Wealth of Nations*," *Journal of Political Economy* 68 (1960): 559.

41. Robert Heilbroner, "The Paradox of Progress: Decline and Decay in the *Wealth of Nations*," in Andrew S. Skinner and Thomas Wilson, eds., *Essays on Adam Smith* (Oxford: Clarendon Press, 1975), pp. 524–539.

42. Ibid., p. 526. (Italics his)

43. *WN* I.ix.13–20.

44. *WN* I.ix.14.

45. *WN* I.ix.20.

46. *WN* I.ix.23.

47. Heilbroner, "The Paradox of Progress," pp. 524–539.

48. Viner, "Adam Smith and Laissez Faire," pp. 134–136.

49. Ibid., p. 134.

50. Ibid., p. 142.

## Conclusion

1. *TMS* II.ii.3.2.

2. See Barry Hindess, *Choice, Rationality, and Social Theory* (London: Unwin Hyman, 1988), pp. 35–41; and Martin Hollis, "Rational Man and Social Science," in R. Harrison, ed., *Rational Action* (Cambridge: Cambridge University Press, 1979), pp. 1–16.

3. Corey Venning, "The World of Adam Smith Revisited," *Studies in Burke and His Time* 19 (1978): 61. Note that Professor Venning does not agree with this reading of Smith.

# BIBLIOGRAPHY


Agassi, Joseph. "Methodological Individualism." *British Journal of Sociology* 26 (1975): 244–270.

Anderson, Gary M. "Mr. Smith and the Preachers: The Economics of Religion in the *Wealth of Nations*." *Journal of Political Economy* 96 (1988): 1066–1088.

Ansbach, Ralph. "The Implications of *The Theory of Moral Sentiments* for Adam Smith's Economic Thought." *History of Political Economy* 4 (1972): 176–206.

Arrow, Kenneth J. "The Division of Labor in the Economy, the Polity, and Society." *Adam Smith and Modern Political Economy*, ed. Gerald P. O'Driscoll, Jr. Ames: Iowa State University Press, 1979: 153–164.

Bagolini, Luigi. "The Topicality of Adam Smith's Notion of Sympathy and Judicial Evaluations." *Essays on Adam Smith*, ed. Andrew S. Skinner and Thomas Wilson. Oxford: Clarendon Press, 1975: 100–113.

Billet, Leonard. "The Just Economy: The Moral Basis for the *Wealth of Nations*." *Review of Social Economy* 34 (1976): 295–315.

Bitterman, H. J. "Adam Smith's Empiricism and the Law of Nature." *Journal of Political Economy* 48 (1940): I: 487–520; II: 703–734.

Bladen, V. W. "Adam Smith on Value." *Essays in Political Economy*, ed. H. A. Innis. Toronto: University of Toronto Press, 1938: 27–43.

Blinder, Alan S. *Hard Heads Soft Hearts*. Reading, Mass.: Addison-Wesley Publishing Company, 1987.

Boulding, Kenneth E. *Adam Smith as an Institutional Economist*. Memphis, Tenn.: P. K. Seidman Foundation, 1976.

Bowley, Marian. "Some Aspects of the Treatment of Capital in the *Wealth of Nations*." *Essays on Adam Smith*, ed. Andrew S. Skinner and Thomas Wilson. Oxford: Clarendon Press, 1975: 361–376.

Brau, Maurice. *Adam Smith's Economics and Its Place in the Development of Political Thought*. London: Croom Helm, 1988.

Brown, E. H. Phelps. "The Labour Market." *Essays on Adam Smith*, ed. Andrew S. Skinner and Thomas Wilson. Oxford: Clarendon Press, 1975: 243–259.

Bryson, Gladys. *Man and Society*. Princeton, N.J.: Princeton University Press, 1945.

Buchanan, James. "The Gauthier Enterprise." *The New Social Contract: Essays on Gauthier*, ed. E. P. Paul, F. D. Miller, Jr., Jeffery Paul, and John Ahrens. New York: Basil Blackwell, 1988: 75–94.

Buchanan, James. "The Justice of Natural Liberty." *Adam Smith and Modern Political Economy*, ed. Gerald P. O'Driscoll, Jr. Ames: Iowa State University Press, 1979: 117–131.

Buchanan, James. "Public Goods and Natural Liberty." *The Market and the State*, ed. Thomas Wilson and Andrew S. Skinner. Oxford: Clarendon Press, 1976: 271–278.

Buckle, H. T. *History of Civilization in England, Vol. II*. London: 1861.

Butler, Joseph. *Fifteen Sermons Preached at Rolls Chapel*. London: Thomas Tegg & Son, 1835.

Campbell, T. D. *Adam Smith's Science of Morals*. London: Allen & Unwin, 1971.

Campbell, William F. "Adam Smith's Theory of Justice, Prudence, and Beneficence." *American Economic Review* 57 (1967): 571–577.

Clay, Marjorie Ann. "'Private Vices, Public Benefits': Adam Smith's Concept of Self-Interest." *Adam Smith and the Wealth of Nations, 1776–1976*, ed. William R. Morrow and Robert E. Stebbins. Richmond, Ky.: Eastern Kentucky University, 1976: 42–61.

Coase, R. H. "Adam Smith's View of Man." Chicago: University of Chicago Graduate School of Business Occasional Paper #50, 1976.

Coats, A. W. "Adam Smith and the Mercantile System." *Essays on Adam Smith*, ed. Andrew S. Skinner and Thomas Wilson. Oxford: Clarendon Press, 1975: 218–236.

Coleman, D. C. "Adam Smith, Businessmen, and the Mercantile System in England." *History of European Ideas*, 9 (1988): 161–170.

Coleman, Jules. "Competition and Cooperation." *Ethics* 97 (1987): 76–90.

Collins, Denis. "Adam Smith's Social Contract." *Business and Professional Ethics Journal* 7 (1988): 119–146.

Cropsey, Joseph. "The Invisible Hand: Moral and Political Considerations." *Adam Smith and Modern Political Economy*, ed. Gerald P. O'Driscoll, Jr. Ames: Iowa State University Press, 1979: 165–176.

Cropsey, Joseph. *Polity and Economy*. Westport, Conn.: Greenwood Press, 1977.

Danner, P. L. "Sympathy and Exchangeable Value: Keys to Adam Smith's Social Philosophy." *Review of Social Economy* 34 (1976): 317–331.

de Vries, Paul H. "Adam Smith's 'Theory' of Justice." *Business and Professional Ethics Journal* 8 (1989):37–56.

Douglas, Paul H. "Smith's Theory of Value and Distribution." *Adam Smith, 1776–1926*, ed. J. M. Clark et al. New York: Augustus M. Kelly, 1926: 77–115.

Ellerman, David. "On the Labor Theory of Property." *Philosophical Forum* 16 (1985): 293–326.

Elster, Jon. "Introduction." *Rational Choice*, ed. Jon Elster. Oxford: Basil Blackwell, 1986: 1–33.

Elzie, L. T. "Self Interest and Economic Power." *Adam Smith and the Wealth of Nations, 1776–1976*, ed. William R. Morrow and Robert E. Stebbins. Richmond, Ky.: Eastern Kentucky University, 1976: 17–30.

Etzioni, Amitai. *The Moral Dimension*. New York: The Free Press, 1988.

Evensky, Jerry. "The Evolution of Adam Smith's Views on Political Economy." *History of Political Economy* 21 (1989): 123–145.

Frank, Robert H. *Passions Within Reason*. New York: W. W. Norton & Company, 1988.

Franklin, Raymond S. "Smithian Economics and Its Pernicious Legacy." *Review of Social Economy* 34 (1976): 379–389.

Friedman, Milton. "Adam Smith's Relevance for 1976." Chicago: University of Chicago Graduate School of Business Occasional Paper #50, 1976.

Gauthier, David. *Morals by Agreement*. Oxford: Clarendon Press, 1986.

Gert, Bernard. "Hobbes' Account of Reason and the Passions." *Thomas Hobbes*, ed. Martin Bertman and Michael LeMalherbe. Paris: Libraire Philosophique J. Vrin, 1989: 89–101.

Gill, Emily. "Justice in Adam Smith: The Right and the Good." *Review of Social Economy* 34 (1976): 275–294.

Ginzberg, Eli. *The House of Adam Smith*. New York: Columbia University Press, 1934.

Goldstein, Leon J. "The Two Theses of Methodological Individualism." *British Journal for the Philosophy of Science* 9 (1958): 1–11.

Goldstein, Maurice. "The Inadequacy of the Principle of Methodological Individualism." *Journal of Philosophy* 53 (1956): 801–813.

Gramm, Warren S. "The Eighteenth Century Adam Smith." Lecture at Ripon College, March 1988.

Gramm, Warren S. "The Selective Interpretation of Adam Smith." *Journal of Economic Issues* 14 (1980): 119–142.

Grampp, William D. "Adam Smith and the Economic Man." *Journal of Political Economy* 56 (1948): 315–336.

Griswald, Charles L., Jr. "Adam Smith on Virtue and Self-Interest." *Journal of Philosophy* 86 (1989): 681–682.

Groenewegen, P. D. "Adam Smith and the Division of Labour: A Bicentenary Estimate." *Australian Economic Papers* 16 (1977): 161–174.

Haakonssen, Knud. *The Science of a Legislator*. Cambridge: Cambridge University Press, 1981.

Hamowy, Ronald. *The Scottish Enlightenment and the Theory of Spontaneous Order*. Carbondale, Ill.: Southern Illinois University Press, 1987.

Hamowy, Ronald. "Adam Smith, Adam Ferguson, and the Division of Labour." *Economica* 35 (1968): 249–259.

Hardin, Garrett. "The Tragedy of the Commons." *Science* 162 (1968): 1243–1248.

Harman, Gilbert. *Moral Agent and Impartial Spectator: The Lindsey Lecture*. Lawrence: University of Kansas Press, 1986.

Hart, David K. "Adam Smith in the Twentieth Century: Is the Vision Intact?" *Exchange* (Winter, 1985): 29–35.

Heilbroner, Robert. *The Essential Adam Smith*. New York: Oxford University Press, 1986.

Heilbroner, Robert. "The Socialization of the Individual in Adam Smith." *History of Political Economy* 14 (1982): 427–439.

Heilbroner, Robert. "The Paradox of Progress: Decline and Decay in *The Wealth of Nations*." *Essays on Adam Smith*, ed. Andrew S. Skinner and Thomas Wilson. Oxford: Clarendon Press, 1975: 524–539.

Hindess, Barry. *Choice, Rationality, and Social Theory*. London: Unwin Hyman, 1988.

Hirsch, Fred. *Social Limits to Growth*. Cambridge, Mass.: Harvard University Press, 1976.

Hirschman, Albert O. *The Passions and the Interests*. Princeton, N.J.: Princeton University Press, 1977.

Hollander, Samuel. "Adam Smith and the Self-Interest Axiom." *Journal of Law and Economics* 20 (1977): 133–152.

Hollis, Martin. "Rational Man and Social Science." *Rational Action*, ed. R. Harrison. Cambridge: Cambridge University Press, 1979: 1–16.

Hont, Istvan, and Michael Ignatieff, ed. *Wealth and Virtue*. Cambridge: Cambridge University Press, 1983.

Hume, David. *A Treatise of Human Nature*, ed. L. A. Selby-Bigge. Oxford: Clarendon Press, 1888; 1960.

Ippolito, Richard A. "The Division of Labor and the Firm." *Economic Inquiry* 15 (1974): 469–492.

Jensen, Hans E. "Sources and Contours of Adam Smith's Conceptualized Reality in the *Wealth of Nations*." *Review of Social Economy* 34 (1976): 259–274.

Kavka, Gregory S. *Hobbesian Moral and Political Theory*. Princeton, N.J.: Princeton University Press, 1986.

Krieger, Leonard. *Kings and Philosophers, 1689–1789*. New York: W. W. Norton & Co., 1970.

Lamb, Robert Boynton. *Property Markets, and the State in Adam Smith's System*. New York: Garland, 1987.

Lamb, Robert Boynton. "Adam Smith's System: Sympathy Not Self-Interest." *Journal of the History of Ideas* 35 (1974): 671–682.

Lerner, Max. "Introduction." Adam Smith, *The Wealth of Nations*, ed. Edwin Cannan. New York: The Modern Library, 1937: v–x.

Levy, David. "David Hume's Invisible Hand in *The Wealth of Nations*: The Public Choice of Moral Information." *Hume Studies* 11 (1985): 110–149.

Levy, David. "Smith and Kant Respond to Mandeville." *History of Political Economy* 14 (1982): 25–37.

Levy, David. "Adam Smith's Natural Law and Social Contract." *Journal of the History of Ideas* 39 (1978): 665–674.

Lewis, Thomas J. "Adam Smith: The Labor Market as the Basis for Natural Right." *Journal of Economic Issues* 11 (1977): 21–50.

Lindgren, J. R. *The Social Philosophy of Adam Smith*. The Hague: Nijhoff, 1973.

Lindgren, J. Ralph. "Adam Smith's Theory of Inquiry." *Journal of Political Economy* 77 (1969): 897–915.

Locke, John. *Two Treatises of Government*, ed. Peter Laslett. Cambridge: Cambridge University Press, 1983.

Lowe, Adolph. "Adam Smith's System of Equilibrium Growth." *Essays on Adam Smith*, ed. Andrew S. Skinner and Thomas Wilson. Oxford: Clarendon Press, 1975: 415–425.

Macfie, A. L. *The Individual in Society: Papers on Adam Smith*. Oxford: Oxford University Press, 1967.

Macfie, A. L. "The Moral Justification of Free Enterprise." *Scottish Journal of Political Economy* (1967): 1–11.

Macfie, A. L. "Adam Smith's Moral Sentiments as Foundation for His *Wealth of Nations*." *Oxford Economic Papers* 11 (1959): 219–228.

Macpherson, C. B. *Democratic Theory*. Oxford: Clarendon Press, 1975.

Macpherson, C. B. *The Political Theory of Possessive Individualism*. Oxford: Oxford University Press, 1972.

Mandelbaum, Maurice. "Societal Laws." *British Journal for the Philosophy of Science* 8 (1957): 211–224.

Mandelbaum, Maurice. "Societal Facts." *British Journal of Sociology* 6 (1955): 305–317.

Mandeville, Bernard. *The Fable of the Bees*, ed. F. B. Kaye. Oxford: Clarendon Press, 1924. Reprinted Indianapolis: Liberty Classics, 1988.

Marx, Karl. *Grundrisse*, ed. and trans. David McLellan. New York: Harper & Row, 1971.

Marx, Karl. *Early Writings*, ed. T. B. Bottomore. New York: McGraw-Hill Book Co., 1963.

Meek, Ronald L., and Andrew S. Skinner. "The Development of Adam Smith's Ideas on the Division of Labour." *Economic Journal* 83 (1973): 1094–1116.

Meek, Ronald L. *Economics and Ideology and Other Essays*. London: Chapman and Hall, Ltd., 1967.

Meek, Ronald L. "The Rise and Fall of the Concept of the Economic Machine." An Inaugural Lecture. Leichester: Leichester University Press, 1965.

Meek, Ronald L. *The Economics of Physiocracy*. London: George Allen and Unwin, Ltd., 1962.

Meek, Ronald L. *Studies in the Labour Theory of Value*. London: Lawrence & Wishart, 1956; 1973.

Mirowski, Philip. "Adam Smith, Empiricism, and the Rate of Profit in Eighteenth Century England." *History of Political Economy* 14 (1982): 178–198.

Morris, Christopher. "The Relation Between Self-Interest and Justice in Contractarian Ethics." *The New Social Contract: Essays on Gauthier*, ed. E. P. Paul, F. D. Miller, Jr., Jeffery Paul, and John Ahrens. New York: Basil Blackwell, 1988: 118–130.

Morrow, Glenn R. "Adam Smith: Moralist and Philosopher." *Adam Smith, 1776–1926*, ed. J. M. Clark et al. New York: Augustus M. Kelly, 1926: pp. 156–179.

Morrow, Glenn R. *The Ethical and Economic Theories of Adam Smith*. New York: Augustus M. Kelley, 1926; 1969.

Morrow, William R., and Robert E. Stebbins, ed. *Adam Smith and the Wealth of Nations, 1776–1976*. Richmond, Ky.: Eastern Kentucky University, 1976.

Musgrave, R. A. "Adam Smith on Public Finance and Distribution," *Essays on Adam Smith*, ed. Andrew S. Skinner and Thomas Wilson. Oxford: Clarendon Press, 1975: 296–319.

Myers, Milton. *The Soul of Modern Economic Man*. Chicago: University of Chicago Press, 1983.

Myrdal, Gunnar. *The Political Element in the Development of Economic Theory*. London: Routledge & Kegan Paul Ltd., 1953.

Napoleoni, Claudio. *Smith, Richardo, Marx*, trans. J. M. A. Gee. New York: John Wiley & Sons, 1975.

Nieli, Russell. "Spheres of Intimacy and the Adam Smith Problem." *Journal of the History of Ideas* 47 (1986): 611–624.

Petrella, Frank. "Individual, Group or Government? Smith, Mill, and Sidgwick." *History of Political Economy* 2 (1970): 152–176.

O'Driscoll, Gerald P., Jr. *Adam Smith and Modern Political Economy.* Ames: Iowa State University Press, 1979.

Quinton, Anthony. "Social Objects." *Proceedings of the Aristotelian Society* 76 (1976): 1-27.

Raphael, D. D. *Adam Smith.* Oxford: Oxford University Press. 1985.

Recktenwald, Horst Claus. "An Adam Smith Renaissance *anno* 1976? The Bicentenary Output—A Reappraisal of His Scholarship." *Journal of Economic Literature* 16 (1978): 56–83.

Rees, Albert. "Compensating Wage Differentials." *Essays on Adam Smith*, ed. Andrew S. Skinner and Thomas Wilson. Oxford: Clarendon Press, 1975: 336–349.

Reisman, D. A. *Adam Smith's Sociological Economics.* London: Croom Helm, 1976.

Richardson, G. B. "Adam Smith on Competition and Increasing Returns." *Essays on Adam Smith*, ed. Andrew S. Skinner and Thomas Wilson. Oxford: Clarendon Press, 1975: 350–360.

Ridpath, John. "Adam Smith and the Founding of Capitalism." Audiotape from Conceptual Conferences, 1988.

Robbins, Lionel. *Political Economy: Past and Present.* London: Macmillan, 1976.

Robbins, Lionel. *The Theory of Economic Policy in English Classical Political Economy.* London: Macmillan, 1952.

Rohrlich, George F. "The Role of Self-Interest in the Social Economy of Life, Liberty, and the Pursuit of Happiness, *Anno* 1976 and Beyond." *Review of Social Economy* 34 (1976): 373–378.

Rosenberg, Nathan. "Adam Smith and Laissez-Faire Revisited." *Adam Smith and Modern Political Economy*, ed. Gerald P. O'Driscoll, Jr. Ames: Iowa State University Press, 1979: 19–34.

Rosenberg, Nathan. "Another Advantage of the Division of Labor." *Journal of Political Economy* 84 (1976): 861–868.

Rosenberg, Nathan. "Adam Smith on Profits—Paradox Lost and Regained." *Essays on Adam Smith*, ed. Andrew S. Skinner and Thomas Wilson. Oxford: Clarendon Press, 1975: 377–389.

Rosenberg, Nathan. "Adam Smith, Consumer Tastes, and Economic Growth." *Journal of Political Economy* 76 (1968): 557–570.

Rosenberg, Nathan. "Adam Smith on the Division of Labour: Two Views or One?" *Economica* 32 (1965): 127–139.

Rosenberg, Nathan. "Some Institutional Aspects of the *Wealth of Nations.*" *Journal of Political Economy* 68 (1960): 361–374.

Samuels, Warren. "The Political Economy of Adam Smith." *Ethics* 87 (1977): 189–207.

Samuels, Warren. "Adam Smith and the Economy as a System of Power." *Review of Social Economy* 31 (1973): 123–137.

Samuels, Warren. *The Classical Theory of Economic Policy*. Cleveland: World Publishing Co., 1966.

Samuelson, Paul. "A Modern Theorist's Vindication of Adam Smith." *American Economic Review* 67 (1977): 42–49.

Saraydon, Edward. "Preferences and Voting Behavior: Smith's Impartial Spectator Revisited." *Economics and Philosophy* 3 (1987): 121–126.

Schneider, Herbert W. *Adam Smith's Moral and Political Philosophy*. New York: Hafner Publishing Co., 1948.

Schneider, Louis. "Adam Smith on Human Nature and Social Circumstance," *Adam Smith and Modern Political Economy*, ed. Gerald P. O'Driscoll, Jr. Ames: Iowa State University Press, 1979: 44–69.

Schumpeter, Joseph. *History of Economic Analysis*, ed. Elizabeth Boody Schumpeter. New York: Oxford University Press, 1954.

Schwartz, Barry. *The Battle for Human Nature*. New York: W. W. Norton & Company, 1986.

Sen, Amartya. *On Ethics and Economics*. Oxford: Basil Blackwell, 1987.

Sen, Amartya. "Adam Smith's Prudence." *Theory and Reality in Development*, ed. S. Lall and F. Stewart. London: Macmillan & Co., 1986: 28–37.

Sen, Amartya. "Rational Fools." *Philosophy and Public Affairs* 6 (1977): 317–344.

Shearmur, Jeremy. "Adam Smith's Second Thoughts: Economic Liberalism and Its Unintended Consequences." *The Adam Smith Club Kirkaldy Paper No. 1*.

Skinner, Andrew S. "A Scottish Contribution to Marxist Sociology?" *Classical and Marxian Political Economy*, ed. Ian Bradley and Michael Howard. New York: St. Martin's Press, 1981.

Skinner, Andrew S. *A System of Social Science: Papers Relating to Adam Smith*. Oxford: Clarendon Press, 1979.

Skinner, Andrew S. and Thomas Wilson. *Essays on Adam Smith*. Oxford: Clarendon Press, 1975.

Smith, Adam. *Lectures on Rhetoric and Belles Lettres*, ed. J. C. Bryce. Oxford: Oxford University Press, 1983.

Smith, Adam. *Essays on Philosophical Subjects*. ed. W. P. D. Wightman and J. C. Bryce. Oxford: Oxford University Press, 1980. Reprinted Indianapolis: Liberty Classics, 1982.

Smith, Adam. *Lectures on Jurisprudence*, ed. R. L. Meek, D. D. Raphael, and P. G. Stein. Oxford: Oxford University Press, 1978.

Smith, Adam. *The Theory of Moral Sentiments*, ed. A. L. Macfie and D. D. Raphael. Oxford: Oxford University Press, 1976. Reprinted Indianapolis: Liberty Classics, 1982.

Smith, Adam. *The Wealth of Nations*, ed. R. H. Campbell and A. S. Skinner. Oxford: Oxford University Press, 1976. Reprinted Indianapolis: Liberty Classics, 1981.

Sobel, Irvin. "Adam Smith: What Kind of Institutionalist Was He?" *Journal of Economic Issues* 13 (1979): 347–368.

Sowell, Thomas. "Adam Smith in Theory and Practice." *Adam Smith and Modern Political Economy*, ed. Gerald P. O'Driscoll, Jr. Ames: Iowa State University Press, 1979: 3–18.

Spengler, Joseph J. "Adam Smith on Human Capital." *American Economic Review* 67 (1977): 32–36.

Spengler, Joseph J. "Adam Smith's Theory of Economic Growth, Part I." *Southern Economic Journal* 21 (1959): 397–415.

Spiegel, H. W. "Adam Smith's Heavenly City." *Adam Smith and Modern Political Economy*, ed. Gerald P. O'Driscoll, Jr. Ames: Iowa State University Press, 1979: 3–18.

Stack, George J. "Self-Interests and Social Value." *Journal of Value Inquiry* 18 (1984): 123–137.

Stein, Peter. "Adam Smith's Jurisprudence—Between Morality and Economics." *Cornell Law Review* 64 (1979): 621–638.

Stigler, George. *Economics or Ethics? Tanner Lectures on Human Values: Volume II*, ed. S. McMurrin. Salt Lake City: University of Utah Press, 1981.

Stigler, George. "Smith's Travels on the Ship of State." *History of Political Economy* 3 (1971): 237–246.

Suttle, B. B. "The Passion of Self-Interest." *American Journal of Economics and Sociology* 46 (1987): 459–472.

Teichgraeber, Richard F., III. *"Free Trade" and Moral Philosophy*. Durham, N.C.: Duke University Press, 1986.

Thompson, Herbert F. "Adam Smith's Philosophy of Science." *Quarterly Journal of Economics* 23 (1965): 212–233.

Tully, John. *A Discourse on Property: John Locke and His Adversaries*. Cambridge: Cambridge University Press, 1980.

Venning, Corey. "The World of Adam Smith Revisited." *Studies in Burke and His Time* 19 (1978): 61–71.

Viner, Jacob. *The Role of Providence in the Social Order*. Philadelphia: American Philosophical Society, 1972.

Viner, Jacob. "Mercantilist Thought." *The International Encyclopedia of the Social Sciences Volume 4*. ed. David L. Sills. New York: Macmillan Company and The Free Press, 1936; 1968: 435–442.

Viner, Jacob. "Adam Smith and Laissez Faire." *Adam Smith, 1776–1926*, ed. J. M. Clark et al. New York: Augustus M. Kelley, 1926: 116–155.

Viner, Jacob. "Guide to John Rae's *Life of Adam Smith*." *Life of Adam Smith*. New York: Augustus M. Kelley, 1895; 1965.

Waszek, Norbert. "Two Concepts of Morality: A Distinction of Adam Smith's Ethics and Its Stoic Origins." *Journal of the History of Ideas* 45 (1984): 591–606.

Watkins, J. W. N. "The Alleged Inadequacy of Methodological Individualism." *Journal of Philosophy* 64 (1957): 390–395.

Watkins, J. W. N. "Historical Explanation in the Social Sciences." *British Journal of the Philosophy of Science* 7 (1957): 104–117.

Waymack, Mark. "Moral Philosophy and Newtonianism in the Scottish Enlightenment." Unpublished dissertation, Johns Hopkins University, 1987.

Werhane, Patricia H. "Freedom, Commodification, and the Alienation of Labor in Adam Smith's Political Economy." *Philosophical Forum*. Forthcoming, 1991.

Werhane, Patricia H. "Adam Smith and His Legacy for Modern Capitalism." *Journal of Philosophy* 86 (1989): 669–680.

West, E. G. "Adam Smith's Economics and Politics." *Adam Smith and Modern Political Economy*, ed. Gerald P. O'Driscoll, Jr. Ames: Iowa State University Press, 1979: 132–152.

West, E. G. "Adam Smith and Alienation: Wealth Increases, Men Decay?" *Essays on Adam Smith*, ed. Andrew S. Skinner and Thomas Wilson. Oxford: Clarendon Press, 1975: 540–552.

West, E. G. "Adam Smith's Two Views on the Division of Labour." *Economica* 31 (1964): 23–32.

West, E. G. "The Political Economy of Alienation: Karl Marx and Adam Smith." *Oxford Economic Papers* 21 (1969): 1–23.

Wicksteed, P. H. *The Common Sense of Political Economy*, ed. Lionel Robbins. London: G. Routledge & Sons, 1946.

Wilson, James O. "Adam Smith on Business Ethics." *California Management Review* 32 (1989): 59–72.

Wilson, Thomas and Andrew Skinner, ed. *The Market and the State*. Oxford: Clarendon Press, 1976.

Wilson, Thomas. "Sympathy and Self-Interest." *The Market and the State*, ed. Thomas Wilson and Andrew Skinner. Oxford: Clarendon Press, 1976: 73–99.

Winch, Donald. *Adam Smith's Politics*. Cambridge: Cambridge University Press, 1978.

Worland, Stephen T. "Mechanistic Analogy and Smith on Exchange." *Review of Social Economy* 34 (1976): 245–257.

# NAME INDEX

# SUBJECT INDEX